ACTIVITIES for
TEACHING STATISTICS
and RESEARCH METHODS

ACTIVITIES for TEACHING STATISTICS and RESEARCH METHODS

A GUIDE FOR PSYCHOLOGY INSTRUCTORS

EDITED BY

Jeffrey R. Stowell and William E. Addison

American Psychological Association • Washington, DC

Published by
American Psychological Association
750 First Street, NE
Washington, DC 20002
www.apa.org

To order
APA Order Department
P.O. Box 92984
Washington, DC 20090-2984
Tel: (800) 374-2721; Direct: (202) 336-5510
Fax: (202) 336-5502; TDD/TTY: (202) 336-6123
Online: www.apa.org/pubs/books
E-mail: order@apa.org

In the U.K., Europe, Africa, and the Middle East, copies may be ordered from
American Psychological Association
3 Henrietta Street
Covent Garden, London
WC2E 8LU England

Typeset in Trump Medieval by Circle Graphics, Inc., Columbia, MD

Printer: Edwards Brothers Malloy, Ann Arbor, MI
Cover Designer: Mercury Publishing Services, Inc., Rockville, MD

The opinions and statements published are the responsibility of the authors, and such opinions and statements do not necessarily represent the policies of the American Psychological Association.

Library of Congress Cataloging-in-Publication Data

Names: Stowell, Jeffrey R., editor. | Addison, William E., editor.
Title: Activities for teaching statistics and research methods : a guide for
 psychology instructors / edited by Jeffrey R. Stowell and William E. Addison.
Description: Washington, DC : American Psychological Association, [2017] |
 Includes bibliographical references and index.
Identifiers: LCCN 2016041081| ISBN 9781433827143 | ISBN 143382714X
Subjects: LCSH: Psychology—Statistical methods. | Psychology—Methodology. |
 Psychology—Study and teaching.
Classification: LCC BF39 .A275 2017 | DDC 150.72/7—dc23
LC record available at https://lccn.loc.gov/2016041081

British Library Cataloguing-in-Publication Data
A CIP record is available from the British Library.

Printed in the United States of America
First Edition

http://dx.doi.org/10.1037/0000024-000

We wish to dedicate this book to our friends and mentors, Dr. Boyd Spencer and Dr. John Best. Boyd Spencer taught statistics at Eastern Illinois University for 37 years, and John Best taught research methods and statistics at Eastern Illinois University for 34 years.

CONTENTS

CONTRIBUTORS

William E. Addison, PhD, Eastern Illinois University, Charleston
Stephanie E. Afful, PhD, Lindenwood University, St. Charles, MO
Bernard C. Beins, PhD, Ithaca College, Ithaca, NY
George Y. Bizer, PhD, Union College, Schenectady, NY
Laura Brandt, MS, College Du Leman, Geneva, Switzerland
Caridad F. Brito, PhD, Eastern Illinois University, Charleston
Andrew N. Christopher, PhD, Albion College, Albion, MI
Dana S. Dunn, PhD, Moravian College, Bethlehem, PA
Sue Frantz, MA, Highline College, Des Moines, WA
Bonnie A. Green, PhD, East Stroudsburg University, East Stroudsburg, PA
Alexis Grosofsky, PhD, Beloit College, Beloit, WI
Thomas E. Heinzen, PhD, William Paterson University, Wayne, NJ
Harold Herzog, PhD, Western Carolina University, Cullowhee, NC
Chris Jones-Cage, PhD, College of the Desert, Palm Desert, CA
Shauna W. Joye, PhD, Georgia Southern University, Statesboro
Mary E. Kite, PhD, Ball State University, Muncie, IN
David S. Kreiner, PhD, University of Central Missouri, Warrensburg
R. Eric Landrum, PhD, Boise State University, Boise, ID
Thomson J. Ling, PhD, Caldwell University, Caldwell, NJ
Robert McEntarffer, PhD, Lincoln Public Schools, Lincoln, NE
Steven A. Myers, PhD, Roosevelt University, Chicago, IL
Thomas P. Pusateri, PhD, Kennesaw State University, Kennesaw, GA
Tamarah Smith, PhD, Cabrini College, Radnor, PA
Jeffrey R. Stowell, PhD, Eastern Illinois University, Charleston
Michael J. Tagler, PhD, Ball State University, Muncie, IN
Christopher L. Thomas, MA, Ball State University, Muncie, IN
Maria Vita, BA, Messiah College, Mechanicsburg, PA
Janie H. Wilson, PhD, Georgia Southern University, Statesboro
Karen Wilson, PhD, St. Francis College, Brooklyn, NY
Joseph A. Wister, PhD, Chatham University, Pittsburgh, PA

ACKNOWLEDGMENTS

Jeff Stowell wishes to thank his wife, Missy, and their six children (Sam, Savannah, Spencer, Sterling, Soren, and Sadie) for their love, patience, and support in everything he does.

William Addison would like to thank his wife, Jayne, and their children (Kaitlin, Will, and James) for their unwavering support over the years. He also wants to thank Jeff Stowell for inviting him to participate in this project—it's always a pleasure to work with him.

ACTIVITIES for
TEACHING STATISTICS
and RESEARCH METHODS

INTRODUCTION

Jeffrey R. Stowell and William E. Addison

Statistics and research methods courses in psychology form the scientific foundation of the discipline and are required in virtually every undergraduate psychology program (Stoloff et al., 2010). Moreover, the American Psychological Association (APA) Board of Educational Affairs Working Group on Strengthening the Common Core of the Introductory Psychology Course (2014) recommended that introductory psychology students should learn scientific reasoning, problem solving, and research methods.

In addition, multiple concepts from these areas are essential components of the Advanced Placement (AP) high school psychology course. The number of students enrolled in AP Psychology continues to rise, as evidenced by significant increases in the number who take the AP Psychology Exam. According to the College Board's (2015) Program Summary Report, more than 275,000 high school students took the AP Psychology Exam in 2015. Approximately 8% to 10% of the AP Psychology Exam is composed of items pertaining to research methods, including descriptive and inferential statistics (The College Board, 2013). This percentage is equal to, or greater than, the percentage for any other topic on the exam.

Although instruction in statistics and research methods is universal in college and AP psychology curricula, these topics are typically among the more challenging for students, in particular for those students who experience math or statistics anxiety (Chew & Dillon, 2014; Chiesi & Primi, 2010; Onwuegbuzie & Wilson, 2003). In addition, comments posted by students on the website http://www.ratemyprofessors.com suggest that statistics courses are generally viewed less favorably than introductory psychology classes (Addison, Stowell, & Reab, 2015). Thus, instructors who teach these topics generally have at least some students with attitudinal barriers to learning the content.

http://dx.doi.org/10.1037/0000024-001
Activities for Teaching Statistics and Research Methods: A Guide for Psychology Instructors, J. R. Stowell and W. E. Addison (Editors)

In many cases, these instructors may consider incorporating innovative strategies for developing and maintaining student interest.

We have learned the pedagogical value of classroom activities through a variety of sources (e.g., attending teaching conferences, reading journal articles, perusing activities books) and have found these activities helpful in our own teaching. After the first author gave a preconference APA workshop that included activities for presenting difficult-to-teach topics in biological psychology, the idea for an activities book shifted to a focus on activities that would benefit a broader audience of instructors. Having worked together on previous research projects, the two of us focused our complementary expertise on this project.

This book was designed to address the need for a comprehensive collection of classroom-tested activities that would engage students, teach correct principles, and inspire instructors. Each chapter describes one or more activities that are pedagogically sound, practical, easily implemented, and effective in helping students learn core topics in statistics and research methods in psychology and, more broadly, the social sciences.

The primary audiences for this book are college instructors who teach undergraduate introductory courses in statistics and research methods (and introductory psychology) and high school AP Psychology teachers. Novice instructors, including graduate students, can use this book as a guided set of activities to launch in their classroom from the start, and veteran instructors can use these activities to replace or supplement existing teaching activities.

The chapter authors represent a diverse group of experts in their respective fields. We have deliberately selected individuals who have distinguished teaching and/or publication records and who are representative of the audience for whom the book is written. Many of the authors have won local or national teaching awards, and some have authored textbooks in the areas of statistics and research methodology. As a whole, the group is demographically, geographically, and institutionally diverse.

The order of the chapters follows the typical progression of topics in an introductory statistics or research methods course. Each chapter is devoted to a concept typically found in these courses and contains one or more classroom activities relevant to this topic.

Each chapter is structured in the following manner:

- *Mini-abstract:* A concise description of the activity
- *Concept:* A brief explanation of the core concept that is being taught with the activity
- *Materials Needed:* Handouts, supplies, equipment, technology, and so on
- *Instructions:* Step-by-step detailed instructions for implementing the activity
- *Assessment:* A description of the methods for assessing student learning (or suggested forms of assessment)
- *Discussion:* A description of expected outcomes, cautions as to what may go wrong, postactivity discussion ideas, follow-up assignments, and other suggestions
- *References:* A list of works cited in the chapter
- *Resources (Appendix):* Handouts for students, experimental stimuli, online resources, and other helpful information

Table 1 includes each chapter's key topics, the expected amount of time needed to complete the activity in class, and the targeted level for the activity. Most of the activities demonstrate introductory-level topics (e.g., measures of central tendency, operational

Table 1 *Chapter Topics, Class Time, and Target Audience*

Chapter no.	Topic	Total class time to complete	AP/ Introductory Psychology	Introduction to Statistics/ Research Methods	Advanced Statistics & Research Methods
1	Reducing statistics anxiety	30 min		x	
2	Interpreting graphs	Two 50-min classes about 1 week apart	x	x	
3	Measures of central tendency	30 min	x	x	
4	Measures of variability	45 min		x	
5	Frequency distributions	45 min	x	x	
6	Normal distribution, percentiles	50 min spread over 3 class periods	x	x	
7	Scatterplots, correlation and regression	30 min		x	
8	Sampling distributions	30 min	x	x	x
9	Hypothesis testing	30 min		x	
10	Confidence intervals	45 min		x	x
11	Type I and II errors	20 min		x	x
12	Statistical power	30 min		x	x
13	Effect sizes, *p* values	25 min		x	x
14	Scientific method	30–45 min	x	x	
15	IVs, DVs, goals of science	20 min for Part 1; additional 5–10 min each for Parts 2–4	x	x	
16	Operational definitions	30 min	x	x	
17	Random assignment	15–20 min to perform the memory experiment (Day 1) and an additional 20–30 min for the full discussion and demonstration of random assignment (Day 2)	x	x	
18	Confounding factors	45 min	x	x	
19	Experimenter bias, participant bias	20 min	x	x	
20	Ethics in human research	30 min	x	x	x
21	Ethics in animal research	50 min	x	x	x
22	Inter-observer reliability	60 min	x	x	
23	Survey construction	Six 50-min class periods		x	x
24	Qualitative research	30–150 min, depending on the scope of use		x	x
25	APA Style writing	35 min/class for peer review of writing early in the term (3 or 4 classes); 70 min/class for such review later in the term (2 or 3 classes)		x	x

Note. IVs = independent variables; DVs = dependent variables; APA = American Psychological Association.

definitions, the scientific method), whereas some topics are more appropriate for advanced undergraduate and/or graduate courses (e.g., confidence intervals, statistical power, qualitative research). Because of space limitations, we could not include a chapter on every important concept in these courses; however, many of the formats used for the activities described in this book (e.g., group activities, research scenarios) can be easily adapted for use with other topics.

We hope that instructors who teach topics in statistics and research methods enjoy using these activities as much as we enjoyed compiling them. We invite readers of this book to share their classroom experiences with us; after all, great teaching ideas deserve to be shared!

REFERENCES

Addison, W. E., Stowell, J. R., & Reab, M. D. (2015). Attributes of introductory psychology and statistics teachers: Findings from comments on RateMyProfessors.com. *Scholarship of Teaching and Learning in Psychology, 1*, 229–234. http://dx.doi.org/10.1037/stl0000034

American Psychological Association. (2014). *Strengthening the common core of the introductory psychology course*. Washington, DC: American Psychological Association, Board of Educational Affairs. Retrieved from http://www.apa.org/ed/governance/bea/intro-psych-report.pdf

Chew, P. K. H., & Dillon, D. B. (2014). Statistics anxiety update: Refining the construct and recommendations for a new research agenda. *Perspectives on Psychological Science, 9*, 196–208. http://dx.doi.org/10.1177/1745691613518077

Chiesi, F., & Primi, C. (2010). Cognitive and non-cognitive factors related to students' statistics achievement. *Statistics Education Research Journal, 9*, 6–26.

The College Board. (2013). *Psychology course description*. Retrieved from http://apcentral.collegeboard.com/apc/public/repository/ap-psychology-course-description.pdf

The College Board. (2015). *Program summary report*. Retrieved from https://secure-media.collegeboard.org/digitalServices/pdf/research/2015/Program-Summary-Report-2015.pdf

Onwuegbuzie, A. J., & Wilson, V. A. (2003). Statistics anxiety: Nature, etiology, antecedents, effects, and treatments: A comprehensive review of the literature. *Teaching in Higher Education, 8*, 195–209. http://dx.doi.org/10.1080/1356251032000052447

Stoloff, M., McCarthy, M., Keller, L., Varfolomeeva, V., Lynch, J., Makara, K., . . . Smiley, W. (2010). The undergraduate psychology major: An examination of structure and sequence. *Teaching of Psychology, 37*, 4–15. http://dx.doi.org/10.1080/00986280903426274

I
STATISTICS

REDUCING ANXIETY IN THE STATISTICS CLASSROOM
Tamarah Smith

Psychology students who experience anxiety in a statistics course are at risk for poor performance in the class. In this chapter I provide a mind-set presentation that helps deconstruct students' misperceptions and reframe their approach to learning statistics. Through this activity, students decrease their statistics anxiety and increase their ability to study and learn statistics.

CONCEPT

The activity described in this chapter is designed to reduce mathematics/statistics anxiety in psychology students enrolled in introductory statistics classes. It is based on the use of cognitive interventions that are designed to restructure beliefs about ability, increase self-confidence, and create a growth mind-set (Paunesku et al., 2015). Research has shown that anxiety about mathematics tends to distract students, thereby reducing their available working memory and hindering their learning and performance (Park, Ramirez, & Beilock, 2014). By using the activity described herein, instructors can help students develop the belief that they are capable of learning statistics, which in turn should reduce their anxiety and improve their learning.

The activity involves teaching students a series of techniques that will help them reduce their anxiety throughout their statistics course and includes a list of empirically based study strategies to help them be more successful in the course. Persuading students to choose to use the techniques that are presented to them requires that they believe the techniques will, in fact, be helpful. To that end, the presentation of these techniques first includes a review of fixed versus growth mind-sets and the relationship between anxiety and academic performance. The goal of the first two parts of the presentation is to help students understand that a growth mind-set exists, that one's ability to be proficient in statistics is not a fixed inherited factor, and that anxiety is not just the result of poor performance but also a cause. Anxiety and mind-set have been shown to have a strong relationship (see Dai & Cromley, 2014), making a mind-set intervention an ideal way to help reduce statistics anxiety. When students believe that they are capable of learning, they will be more likely to use the anxiety reduction and study strategies described at the end of the presentation. These techniques should allow them to reduce their anxiety and succeed in the course (see Blackwell, Trzesniewski, & Dweck, 2007).

http://dx.doi.org/10.1037/0000024-002
Activities for Teaching Statistics and Research Methods: A Guide for Psychology Instructors, J. R. Stowell and W. E. Addison (Editors)

On the first day of class, students complete six items related to implicit theories of intelligence and school performance (Wang & Ng, 2012), the first 23 items of the Statistics Anxiety Rating Scale (Cruise, Cash, & Bolton, 1985), and an 18-item assessment of statistics knowledge (Smith, 2015; available from the author upon request). The activity takes the form of a visual presentation with the main points highlighted in a bulleted fashion (see the Instructions section of this chapter). Free alternative measures can be accessed on the Assessment Research Tools for Improving Statistical Thinking website (see University of Minnesota, 2006).

After students complete the baseline assessment the instructor should present the four topics below, in order. Students should be asked to take notes during the presentation, which takes approximately 30 minutes to complete (including the viewing of the video mentioned below).

1. *Myths about math.* The instructor should address the myth that people are born better at either math or reading/writing and then introduce the concept of mind-set. The instructor should emphasize the power of our beliefs and that practice can improve performance regardless of our starting point or past experiences (see http://www.carnegiefoundation.org). For an example of the power of belief, see Tomasetto, Matteucci, Carugati, and Selleri's (2009) study, which showed that students who identify as being better with language than math can do better on a math test than students who identify as being better at math than language.
2. *Brain activity and mind-set.* Briceno's (2012) video suggests that people with a growth mind-set subscribe to the idea that they can learn through practice, whereas a fixed mind-set reflects beliefs of innate, and therefore fixed, intelligence. Research using brain scans has shown that after completing a task, individuals with a growth mind-set had increased brain activity associated with attention errors, which in turn led to increased accuracy. This type of brain activity was not present in individuals with a fixed mind-set, which resulted in reduced accuracy due to less attention to errors (Moser, Schroder, Heeter, Moran, & Lee, 2011).
3. *Attitudes and anxiety.* The instructor should explain to students that their attitudes, including their beliefs about their abilities and past math experiences, can cause anxiety. This anxiety taxes their working memory, which detracts from their ability to focus on learning statistics (see Park et al., 2014).
4. *Evidence-based, concrete ways to manage anxiety and learn.* This final section provides concrete strategies that students can use to reduce or reappraise their anxiety and maximize their efforts when studying (see Appendix 1.1). Knowledge of these strategies is important for empowering students to enhance their skills rather than feeling they are fixed where they are, with no way to change. In this section, the instructor describes two techniques for avoiding negative consequences of anxiety: (a) a self-regulation model for reducing anxiety and (b) an anxiety reappraisal technique. These techniques ask students to identify their awareness of their feelings prior to the beginning of the course and show students how to use the strategies they have learned on their own throughout the course. While completing their coursework, students should

take steps to help reduce or reappraise their anxiety (e.g., take a break, stretch, use deep breathing), and they should reflect afterward on what did and did not work for them. They should acknowledge and commend themselves for any progress they make. The instructor's presentation should include attention to problems with instructionalism and the need for active learning (see Mueller & Oppenheimer, 2014, for an example). Effective modes of learning include using what you know before you start, connecting different topics, being an active learner, considering your environment, and pacing your work. Sawyer (2006) provided more information in this area.

ASSESSMENT   To determine whether anxiety has been reduced and knowledge increased, students complete pre- and postassessments using the three instruments listed in the Materials section. Two instruments measure the main outcomes: anxiety toward statistics (Statistics Anxiety Rating Scale) and knowledge of statistics (Research Methods Skills Assessment; Smith, 2015). Given that the theoretical basis for the activity incorporates theories of intelligence, a third set of questions (Wang & Ng, 2012) measures students' beliefs of implicit intelligence to determine the extent to which the presentation was effective in restructuring their beliefs. Again, all instruments should be distributed before and after the presentation, in counterbalanced packets.

Student names or identification numbers should be recorded on each packet to allow for pre and post comparisons, but students should be ensured that the materials are confidential. The second assessment should be given during the last week of the course and follow the same procedure as the first assessment. In regard to comparisons of pre and post scores, authors of past studies who have used techniques similar to the one described here have observed large effect sizes ($ds > .60$) for both anxiety and beliefs about intelligence (Blackwell et al., 2007).

DISCUSSION   The activity I have described in this chapter provides a general framework for reducing anxiety. Some instructors present students with this type of material throughout the duration of a course (Karimi & Venkatesan, 2009). Other activities designed to reduce anxiety include showing documentaries (Hekimoglu & Kittrell, 2010), or having students write about how they feel prior to an examination, which has been found to reduce anxiety in educational settings (Park et al., 2014). Instructors may wish to use these techniques to supplement the activity described here. This activity focuses solely on the students; however, instructors can benefit from incorporating productive feedback to help support students maintain continued mind-set growth (Yeager et al., 2014).

Instructors should exercise caution when working with students who may have especially high levels of anxiety. It is possible that class discussions about anxiety could trigger or heighten fears to an unhealthy level in some students. Instructors should be aware of this possibility and provide appropriate assistance, such as directing students who show significant signs of distress to the institution's counseling center.

REFERENCES   Blackwell, L. S., Trzesniewski, K. H., & Dweck, C. S. (2007). Implicit theories of intelligence predict achievement across an adolescent transition: A longitudinal study and an intervention. *Child Development*, 78, 246–263. http://dx.doi.org/10.1111/j.1467-8624.2007.00995.x

Briceno, E. (2012, November). *The power of belief: Mindset and success* [TED talk]. Retrieved from https://www.youtube.com/watch?v=pN34FNbOKXc

Cruise, R. J., Cash, R. W., & Bolton, D. L. (1985, August). *The development and validation of an instrument to measure statistical anxiety.* Paper presented at the annual meeting of the American Educational Research Association, Chicago, IL.

Dai, T., & Cromley, J. G. (2014). Changes in implicit theories of ability in biology and dropout from STEM majors: A latent growth curve approach. *Contemporary Educational Psychology, 39,* 233–247. http://dx.doi.org/10.1016/j.cedpsych.2014.06.003

Hekimoglu, S., & Kittrell, E. (2010). Challenging students' beliefs about mathematics: The use of documentary to alter perceptions of efficacy. *PRIMUS: Problems, Resources, and Issues in Mathematics Undergraduate Studies, 20,* 299–331. http://dx.doi.org/10.1080/10511970802293956

Karimi, A., & Venkatesan, S. (2009). Cognitive behavior group therapy in mathematics anxiety. *Journal of the Indian Academy of Applied Psychology, 35,* 299–303.

Moser, J. S., Schroder, H. S., Heeter, C., Moran, T. P., & Lee, Y.-H. (2011). Mind your errors: Evidence for a neural mechanism linking growth mind-set to adaptive posterror adjustments. *Psychological Science, 22,* 1484–1489. http://dx.doi.org/10.1177/0956797611419520

Mueller, P. A., & Oppenheimer, D. M. (2014). The pen is mightier than the keyboard: Advantages of longhand over laptop note taking. *Psychological Science, 25,* 1159–1168. http://dx.doi.org/10.1177/0956797614524581

Park, D., Ramirez, G., & Beilock, S. L. (2014). The role of expressive writing in math anxiety. *Journal of Experimental Psychology: Applied, 20,* 103–111. http://dx.doi.org/10.1037/xap0000013

Paunesku, D., Walton, G. M., Romero, C., Smith, E. N., Yeager, D. S., & Dweck, C. S. (2015). Mind-set interventions are a scalable treatment for academic underachievement. *Psychological Science, 26,* 784–793. http://dx.doi.org/10.1177/0956797615571017

Sawyer, R. K. (Ed.). (2006). The new science of learning. In *The Cambridge Handbook of the Learning Sciences* (pp. 1–16). New York: Cambridge University Press.

Smith, T. (2015, August). *Reliability and validity of the Research Methods Skill Assessment.* Poster presented at the 122nd Annual Convention of the American Psychological Association, Toronto, Ontario, Canada.

Tomasetto, C., Matteucci, M. C., Carugati, F., & Selleri, P. (2009). Effect of task presentation on students' performances in introductory statistics courses. *Social Psychology of Education, 12,* 191–211. http://dx.doi.org/10.1007/s11218-008-9081-z

University of Minnesota. (2006, June). *Assessment resource tools for improving statistical thinking.* Retrieved from https://apps3.cehd.umn.edu/artist/resources.html

Wang, Q., & Ng, F. F. Y. (2012). Chinese students' implicit theories of intelligence and school performance: Implications for their approach to schoolwork. *Personality and Individual Differences, 52,* 930–935. http://dx.doi.org/10.1016/j.paid.2012.01.024

Yeager, D. S., Purdie-Vaughns, V., Garcia, J., Apfel, N., Brzustoski, P., Master, A., . . . Cohen, G. L. (2014). Breaking the cycle of mistrust: Wise interventions to provide critical feedback across the racial divide. *Journal of Experimental Psychology: General, 143,* 804–824. http://dx.doi.org/10.1037/a0033906

Appendix 1.1

Evidence-Based, Concrete Ways to
Manage Anxiety and Learn

- **Use evidence-based study techniques.** No one is going to open up your head and drop knowledge into it. You have to work for it, and your effort, not natural talent, is what helps you learn, by building neural networks in your brain. Try the following techniques:
  - Be an active learner by writing things down, drawing ideas, diagramming them, and even saying them aloud.
  - Use what you know before you start studying to help you connect the new material to other topics.
  - Use the tools, people, and environment you are in to help you apply content.
  - Pace and space! Consider how athletes train. They do not just practice the night before a game; they practice every day.
- **Manage your anxiety.** It is common for people to experience some degree of nervousness when they are faced with a new task. You may not be able to stop anxiety from coming, but you can stop it from managing you. Self-regulation is key! Consider the following steps while you are working on a project that causes you anxiety.
  - *Before you start your work.* Think about how you feel. Are you about to do something you feel is going to be difficult? If so, then you know to be on the lookout for anxiety. Be aware that the physiological arousal caused by anxiety can actually aid performance.
  - *While you are working.* If you begin to feel anxious, remind yourself that you are capable of learning. Take a time-out if you need it; take a walk, joke with a friend. Then reconsider what it is you are trying to do and try to identify the parts you are not understanding. Remember, arousal from anxiety does not have to be a bad thing; it can actually help you perform better.
  - *Reflect.* How do you feel when you are done? What were the strategies that helped you get the work done? Congratulate yourself. Even if the work is not perfect or complete, it is more than what you had when you started. Good for you!

2 HOW TO LIE WITH THE Y-AXIS
Thomas E. Heinzen

This activity teaches students how to become critical consumers of graphic data by teaching them how to use the y-axis to create false or misleading visual impressions of data.

CONCEPT

Darrell Huff's (1954) classic book *How to Lie With Statistics* has been a popular, helpful book for many years, and deservedly so. Huff described a number of different ways to lie with statistics and devoted an entire chapter (Chapter 5, "The Gee-Whiz Graph") to lying with graphs. Huff understood that teaching people how to lie with statistics is a bit like creating a "manual for swindlers." His counterargument, which has held sway across more than 500,000 copies of his book, is that the "crooks already know these tricks; honest men must learn them in self-defense" (p. 9).

Edward Tufte (2001) used a different approach to teach about the importance of graphical design. He demonstrated how the 1986 Challenger space shuttle tragedy—the spacecraft failed because the O-rings became less flexible in cold temperatures—may well have been averted by a properly constructed graph. NASA decision makers had requested a visual summary of data about temperature and damage to the O-rings the night before the launch. The engineers, however, submitted a graph organized by time (when they obtained the data) rather than by temperature. Hidden from view was a clear message: The O-rings will fail in cold temperatures: DO NOT LAUNCH. The next morning, icicles hung from the infrastructure, a fiery plume appeared at ignition, and the Challenger, already 9 miles above earth, exploded 73 seconds after launch.

The first outcome of the exercise described in this chapter will be that students know how to use current software (e.g., Microsoft Excel) to create a truth-telling graph. The second outcome is that students will know when someone is subtly mislabeling the y-axis and using the visual display to give a misleading impression.

MATERIALS NEEDED

1. Part 1 requires asking students to bring to class two or three graphs from magazines, newspapers, or online sources. Ask them to evaluate the graphs using the following questions as a guide: What is being measured? What is being compared? What is not being compared that should be compared? Does it appear that the graph's creator is deliberately trying to mislead the viewer? For example, an automobile manufacturer might create a graph that compares the percentage of its brand of vehicles still on the road after 10 years to other brands—but leave out comparison brands that perform better than their own.

http://dx.doi.org/10.1037/0000024-003

Activities for Teaching Statistics and Research Methods: A Guide for Psychology Instructors, J. R. Stowell and W. E. Addison (Editors)

2. Part 2 requires the use of Microsoft Excel software. The graphing exercise language is general enough that it should allow professors and students to create visual displays of data regardless of the Excel version they are using.

Even though there are many ways to lie with graphs, the most common techniques involve the vertical *y*-axis, which tells us what the graph is measuring. For example, the graphs created in this activity represent data from an annual report about how many people are taking advantage of a sponsored professional development program (see Figures 2.1 and 2.2). The purpose of the annual report is, of course, to get more money from the funders by convincing them that they have a growing number of participants in the program. The two graphs are based on the same data, so the lies are subtle. For example, in

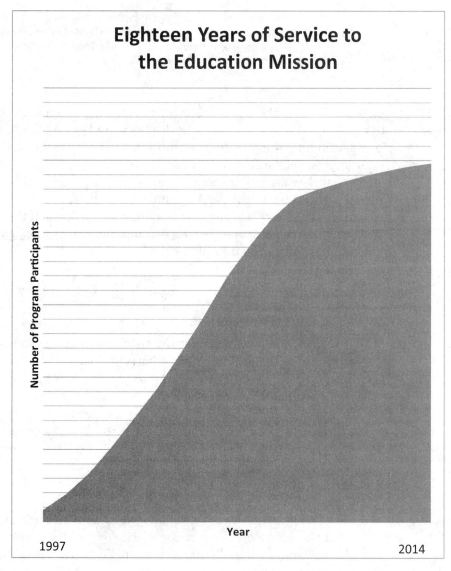

Figure 2.1.   A cumulative frequency graph that uses vague information to leave a false impression that a professional development program is still growing.

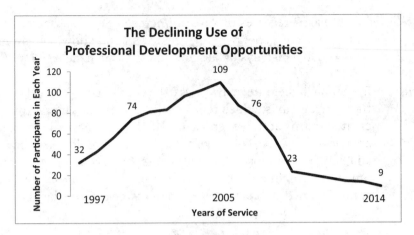

Figure 2.2. An annual frequency line graph. The y-axis on this graph displays a range just below the lowest number and just above the highest number. The label for the y-axis is worded more clearly. The horizontal lines have been removed. And it tells an honest story: This program is finished!

Figure 2.1, labeling the y-axis "Number of program participants" is not an outright lie, but a more precise description of the display would be "A cumulative frequency graph showing the total number of program participants." The graph can still leave a false visual impression, but the more descriptive and accurate label would inform careful readers that they are looking at a visual deception. A cumulative frequency distribution simply adds up all the participants, year by year, so that it is impossible for the line on this graph (representing a total number) to go down. If there were no additional participants after a particular year, then a cumulative frequency graph would still show a straight, horizontal line from that point.

Part 1: Create an Area Graph From the Cumulative Frequency Distribution

1. In an empty Microsoft Excel Workbook, copy the data set from Table 2.1.
2. Highlight the numbers from the right-most column only (the cumulative frequency distribution) and go to "Charts," which appears under the "Insert" menu at the top. Click on "Charts" and select the "Area" chart to represent the data. (Most graphics programs put the most commonly used and appropriate type of graph in the upper left corner.) In the newly created chart, the numbers 1 through 18 along the bottom represent the 18 years of data points (1997–2014). For this activity, it is not necessary to label the years from 1997 on the left to 2014 on the far right. Selecting other types of charts will represent the same data in different ways and may be instructive depending on the data set and instructional focus.
3. Chart elements can easily be changed by right-clicking on a particular section of the graph. Right click on the y-axis to label the vertical y-axis "Number of Participants." Select "Format Axis" to do lots of other things, such as change the scale (minimum and maximum values of the y-axis) and to rotate the label "Number of Participants" to a horizontal position so that it is easier to read.

Table 2.1 *Frequency Distribution of Participants by Year*

Year	Frequency distribution	Cumulative frequency distribution
1997	32	32
1998	43	75
1999	57	132
2000	74	206
2001	81	287
2002	83	370
2003	96	466
2004	102	568
2005	109	677
2006	87	764
2007	76	840
2008	56	896
2009	23	919
2010	20	939
2011	17	956
2012	14	970
2013	13	983
2014	9	992

4. Eventually, the area chart should look something like Figure 2.1. The design of this graph conveys the false impression that the professional development program is continuing to grow even though, in reality, it is serving fewer new people each year. In a graph based on cumulative frequency data a flat line means that no new people participated in the program.

Graphing the cumulative number of participants could easily lead the viewer to reach the conclusion that the professional development program continued to grow over the course of 18 years, even though the annual number of new participants was declining. This graph is called an *area graph* and is based on the cumulative frequency distribution.

Part 2: Create a Line Graph From the Frequency Distribution

Repeat all the steps in Part 1, except this time highlight the data column with the heading "Frequency Distribution." Now, label your title and your *y*-axis more specifically (i.e., specify that it represents the number of participants *each year*). Make your graph appear flatter by stretching the graph horizontally. Eventually, your visual display should look something like Figure 2.2. Also, notice what happens to the visual impression of the data when you change the range of the *y*-axis to a maximum of 1,000.

You cannot design a proper *y*-axis without considering the horizontal *x*-axis. In this case, the default for Excel can include every year or only a few years. When you have many years, as we do in this graph, the *x*-axis tends to get a little crowded. Tufte's (2001) recommendation for this kind of situation is to follow the principle of the data-to-ink ratio: more data/less ink. The most relevant year for this graph, in addition to the end-point years of 1997 and 2014, is the peak year of 2005, when the number of participants

Table 2.2 *A Summary of How to Lie and Tell the Truth Using the Y-Axis*

A liar's guide to the y-axis	A truth-teller's guide to the y-axis
Change the range of the y-axis to exaggerate or restrict differences in the data.	Set an appropriate y-axis scale (e.g., use 0 as the lower limit) so that it accurately represents the data story.
Describe what is being measured on the y-axis with vague, general wording.	Be specific when labeling what is being measured.
Distract people with pictures, extra lines, and redundant information. Use gimmicks such as three-dimensional images, spinning photographs, or anything that distracts from the bad news that you do not want to tell.	Avoid any distracting images that draw attention away from the data.

began to decline. The data:ink ratio encourages one to leave out any marks or information that are not strictly necessary. Graphs should not be viewed as an opportunity to show off your graphical creativity; they are an opportunity to show off your scientific creativity by "letting the data speak."

Figures 2.1 and 2.2 are based on the same data, yet they look very different and appear to tell different data stories. It is easy to lie using the y-axis, and recognizing these tricks can earn you a reputation as a particularly useful employee. As a psychology major, future employers may expect to take advantage of your training in statistics by asking you to summarize sales figures, evaluate a new training program, summarize the performance appraisals of new hires, or document how many people are taking advantage of professional development opportunities. In this case, the insight in Figure 2.2 that is not apparent in Figure 2.1 is that fewer people were using the professional development program after 2005; by 2014, only nine were taking advantage of the program. This insight prompts better questions, such as "What started to happen after 2005 that may have discouraged employees from taking advantage of professional development opportunities?" Table 2.2 contains a summary of how to lie and tell the truth using the y-axis.

ASSESSMENT

The most effective way to assess learning may be by presenting students with the graphs they first brought to class. Have them evaluate those graphs again, identify what they have learned from the exercise, and propose practical improvements to their graphs that will improve interpretation. A checklist, such as the one below, might be handy.

- How could the label for the y-axis be made more specific?
- Is the label for the y-axis rotated horizontally so that it is easy to read?
- What would the visual display look like if the range were flattened or stretched?
- What would the visual display look like if the borders of the graph were flattened or stretched?
- Does it appear that the graph's creator is deliberately trying to mislead the viewer?
- Are there any unnecessary lines or images that distract from the data story?

REFERENCES

Huff, D. (1954). *How to lie with statistics.* New York, NY: W. W. Norton.

Tufte, E. (2001). *The visual display of quantitative information.* Cheshire, CT: Graphics Press.

3 SUMMARIZING DATA USING MEASURES OF CENTRAL TENDENCY: A GROUP ACTIVITY

Thomson J. Ling

This activity is designed to help students understand the three measures of central tendency. Students, working in small groups, calculate various summary statistics of their height to understand the concepts of mean, median, and mode.

CONCEPT

First-year college students spend an average of 14 hours studying per week (Center for Postsecondary Research, 2015). Teenagers send and receive an average of 30 text messages per day (Pew Research Center, 2015). These numbers summarize the data collected from a large number of individuals.

When collecting data from a group, it is often helpful to summarize that information using one or more commonly accepted descriptive statistics. For example, suppose you collected data from a particular class on the number of hours each student had slept the previous night. You could create a spreadsheet listing the name and the number of hours slept for every student in the class. Although this spreadsheet would indicate how long each student slept, it would not be helpful for understanding the class as a whole. To better understand the entire data set, you would need to summarize all of this information into a single, manageable number that is most typical or best represents the entire class. For example, you could calculate the average hours slept or another number that represents the typical student in the class. An average or typical score for a group of data is known in statistics as a measure of *central tendency*.

Three common statistics are used to represent central tendency: (a) the mode, (b) the mean, and (c) the median. The *mode* is the most frequently occurring value, or the most commonly represented number. A group could have one mode, more than one mode, or no mode. The *mean* is what you get by summing the scores and dividing this amount by the number of scores (i.e., the arithmetic average). One challenge with using the mean to represent a group is that it can shift noticeably when the data include outliers. Finally, the *median* is the number that divides the group into equal halves, with half of the scores above the median and the other half below the median. Although the median is still vulnerable to outliers, it is less likely than the mean to yield a misleading number, which is why it is preferred for skewed distributions.

MATERIALS NEEDED

Each student will need a calculator, a pen or pencil, and a copy of the tally sheet I have created for this activity (see Appendix 3.1). You as the instructor should have enough tape measures so that students can share (I recommend one tape measure for every two students).

http://dx.doi.org/10.1037/0000024-004
Activities for Teaching Statistics and Research Methods: A Guide for Psychology Instructors, J. R. Stowell and W. E. Addison (Editors)

Because this activity is designed to give students hands-on experience with the measures of central tendency, it is best implemented after explanations of the mean, median, and mode are provided. You should explain to students that they will be working in small groups (i.e., three to five students) to calculate statistics representing central tendency to summarize their group members' heights. Students will calculate the mean and median height for their own group members and then team up with a second group to calculate modal height.

Arrange students into small groups and provide each student with a tally sheet (see Appendix 3.1). Each group will need at least one tape measure. In the "Data" section of the tally sheet, ask students to record the name and height (in inches) of each student in their group, using the tape measure.

Mean

Students in each group should record the sum of the heights of all members of their group under the "Mean" section of the tally sheet. Students can then calculate the mean height for their group by dividing the sum of the heights by the number of students in the group.

Median

In their groups, students should arrange themselves from shortest to tallest by standing next to one another. Under the "Median" section of the tally sheet, students should record the group members in order of ascending height (i.e., shortest to tallest) and record the height of each person. The median height is the middle height that divides the group in two. For example, with five students in a group, the median would be the third-shortest (or third-tallest) height. If students are in an even-numbered group, students should calculate the mean of the two middle heights to obtain the median height.

Mode

Once students have found the mean and median height for their group, have them work with a second group to calculate the modal height of the groups. The true modal height would be based on raw scores, but for demonstration purposes it is easier to find a mode among a small number of scores if the scores are grouped into categories first. Under the "Mode" section of the tally sheet there is a blank bar graph with six height categories (under 60", 60"–under 63", 63"–under 66", 66"–under 69", 69"–under 72", 72" and over). In the combined group, students should determine into which height category each student falls, write that student's name in a blank box above the height category label, and shade in the blank box. After all students have recorded their names in a box, they should find the height category with the most boxes shaded to determine the modal height for the group.

Impact of Extreme Scores

After students have calculated the mean, median, and mode for their group, ask them to think about what would happen to each of these measures if an extremely tall person (e.g., 7 feet) joined their group. They should then be asked to recalculate the mean, median, and mode while taking this additional person into account. This activity can be helpful in illustrating how the mean is the most affected by extreme scores and the mode is least affected.

ASSESSMENT   To assess their understanding of the measures of central tendency, students can be asked to define the mean, median, and mode. Furthermore, they can be asked to consider other scenarios in which concepts of central tendency would be beneficial in summarizing information.

To assess their understanding of measures of central tendency, students can be asked to calculate the mean, median, or mode for other, common distributions. The following are some examples:

- the number of points per player in a basketball game,
- the number of commercials during a commercial break,
- the number of miles driven daily by family members,
- the number of "likes" on a social media post, and
- fictional data for scores on an exam.

DISCUSSION   Students commonly report that the activity described in this chapter enhances their understanding of the measures of central tendency. You can discuss with your classes how these measures simplify how psychology professionals communicate information about data to others and how they are able to summarize a large amount of data with single measures of central tendency.

Be prepared to discuss instances in which two or more modes are present or when no mode is present. For this reason, I recommend that histograms be introduced in conjunction with the concept of central tendency. In addition, the activity presented in this chapter can be modified to help students connect measurement scales with central tendency. For example, instructors can explain that whereas height (a ratio scale) lends itself well to calculations of the mean, median, and mode, student gender (a nominal scale) can be summarized only with the mode. You may also wish to challenge your students by asking them to calculate the mean, median, or mode with data from the entire class as opposed to using the data from a small group of students.

REFERENCES

Center for Postsecondary Research. (2015). *National Survey of Student Engagement: Summary tables*. Retrieved from http://nsse.indiana.edu/html/summary_tables.cfm

Pew Research Center. (2015). *Teens, social media & technology overview 2015*. Retrieved from http://www.pewinternet.org/files/2015/04/PI_TeensandTech_Update2015_0409151.pdf

Appendix 3.1

Tally Sheet for Central Tendency Activity

Data

Student name	Height (in inches)

Mean

Sum the heights from your group	Divide by the number of students in your group (3, 4, or 5)	Mean height
	÷ 3, ÷ 4, or ÷ 5	

Median

Shortest person's height	Next-shortest person's height (this box not used with groups of 3)	Median height (this box is a calculated number with groups of 4)	Second-tallest person's height (this box not used with groups of 3)	Tallest person's height

Mode

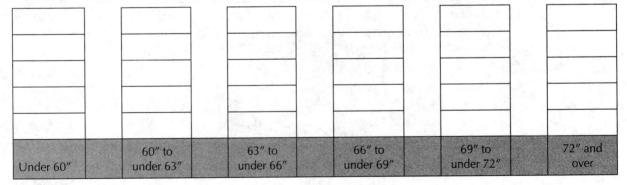

Under 60"	60" to under 63"	63" to under 66"	66" to under 69"	69" to under 72"	72" and over

4   HOW FAST IS YOUR INTERNET?
AN ACTIVITY FOR TEACHING VARIANCE
AND STANDARD DEVIATION

Bonnie A. Green and Jeffrey R. Stowell

The two activities described in this chapter were designed to help students understand the concepts of variance and standard deviation. The first activity focuses on the comprehension and calculation of these measures, and the second activity requires students to visualize how changes in distributions of scores affect the value of the standard deviation.

CONCEPT

The concept of variability is central to understanding all statistics, and deviation is a key concept in understanding variability. The deviation tells us about an individual score relative to other scores in the distribution. If you were interested in knowing the average amount of deviation among a group of scores as a measure of the variability, or "spread-out-ness," of the data, you could sum each score's deviation from the mean, $\Sigma(X - \bar{X})$, and divide by the total number of observations, N. However, the result always equals 0, proving meaningless. Students can more easily see the problem when the data are presented as shown in Table 4.1.

Variance circumvents the problem that the sum of the deviations from the mean always equals 0, by squaring the deviations from the mean before dividing by the total number of observations:

$$\text{Variance} = \frac{\Sigma(X - \bar{X})^2}{N}.$$

However, squaring the deviations also squares the unit of measurement, leaving us with a conceptual challenge students should be able to identify. Thus, with a little guidance from you, the professor, students readily figure out that we can just take the square root of the variance, which results in a measure of variability in the original (unsquared) units, that is, the standard deviation (*SD*):

$$SD = \sqrt{\frac{\Sigma(X - \bar{X})^2}{N - 1}}.$$

http://dx.doi.org/10.1037/0000024-005

Activities for Teaching Statistics and Research Methods: A Guide for Psychology Instructors, J. R. Stowell and W. E. Addison (Editors)

Table 4.1   *Scores, Deviations, and Squared Deviations*

X	$X - \bar{X} =$ deviation	Squared deviation
2	$2 - 3 = -1$	1
5	$5 - 3 = 2$	4
1	$1 - 3 = -2$	4
4	$4 - 3 = 1$	1
3	$3 - 3 = 0$	0
$\bar{X} = 3$	$\Sigma(X - \bar{X}) = 0$	$\Sigma(X - \bar{X})^2 = 10$

Two things should be noted about the formulas presented above:

1. These are not computational formulas; they are definitional formulas, but they should help students see how deviations are built into the formula.
2. Population and sample symbols (e.g., σ and s) should not be used until after students have been taught the difference between sample and population statistics, biased estimators, and the differences in formulas (i.e., the population *SD* uses N, and the inferential *SD* uses $N - 1$ as the denominator).

MATERIALS NEEDED

Prior to the class in which the first activity is conducted, students who own a mobile device (e.g., smartphone, tablet, laptop) should download a free Internet connection speed app (e.g., Ookla speedtest; see http://speedtest.net) or be prepared to connect to another Internet connection speed test website using a browser on their mobile device. According to a recent report, the large majority (85%) of college-age adults own a smartphone (Smith, 2015).

Instructions for Activity No. 1: Internet Speed

Assign students to groups of four to six members, with one member of each group designated as the recorder. If fewer than three group members have a smartphone, you may need to rearrange or combine the groups until there are at least three members who do. Ask each person with a mobile device to verify his or her device's Internet connection and to open the connection speed app (or enter the URL of the Internet connection speed website). At a signal from the instructor, all students should begin the Internet speed test. After completing the test, each group's recorder should write down the group members' names, type of data connection (e.g., Wi-Fi, 2G, 3G, 4G, 4G LTE) and download speed (in megabits per second [Mbps]). After data collection, have each group calculate the group's mean, deviations from the mean, squared deviations, variance, and *SD* of their download speeds. In addition, or alternatively, students can repeat the Internet-speed test five times at different locations or at different times of the day to generate comparison data using a within-subject design. First as groups, and then as a class, the following questions can be addressed:

- What are some factors that influence the Internet speed of different devices?
- What are some factors that influence the Internet speed of the same device over time?

- If every student in the class had the same mobile device and data connection, how would this affect measures of variability?
- If you had a choice between two data connection plans that advertised download speeds of 50 Mbps, but one had data speeds with an *SD* of 25 Mbps, and the other plan had data speeds with an *SD* of 5 Mbps, which plan would you choose, and why?
- If you were to calculate Internet speed variability using data from all groups, how might the class's variance and *SD* differ from each group's values?
- If students were to calculate the variance and *SD* across members of a group and within the same individual over time, which set of data would produce larger measures of variability? Why?

Instructions for Activity No. 2: Interactive Data Visualization

The second activity requires a free download and installation of the Wolfram CDF Player at http://demonstrations.wolfram.com/download-cdf-player.html. The installation file is nearly 700 MB and requires about 2.5 GB of disk space. Links to download the interactive data visualizations are included in the References (i.e., McClelland, 2015; Wolfram, 2015).

The first data visualization demonstration, *Mean and Standard Deviation of a Distribution* (Wolfram, 2015), shows a hypothetical distribution of scores, which for this activity can represent the class's possible distribution of Internet speeds. To begin, open and display the demonstration on a presentation computer. In this visualization the vertical red bar represents the mean of the distribution, and the shaded blue region on either side of the vertical red bar represents the *SD* (see Wolfram, 2015). Ask students to predict how the *SD* would change from the default value if you create a skewed distribution with a large proportion of high Internet speeds and a small proportion of slow Internet speeds (i.e., a negatively skewed distribution). Ask a student to come to the presentation computer and drag each of the data points up or down on the chart until the distribution resembles roughly the scenario you just described. The resulting chart should look similar to the one in Figure 4.1. Of course, scenarios that produce other distributions of Internet speeds (or other real-life variables) are possible and should be presented so that students can gain additional practice and understanding.

The second data visualization demonstration, *The Normal Distribution* (see McClelland, 2015), shows a standard normal curve drawn in red and a second normal curve drawn in blue with two parameters you can manipulate: the mean and *SD*. Suggest to students that the fixed distribution (standard normal curve) represents a normal distribution of data speeds from 2G (slower) mobile devices. Ask a student to manipulate on the presentation computer either or both of the parameters to represent how data speeds from a sample of 4G LTE (faster) mobile devices might compare. Figure 4.2 illustrates a case in which 4G LTE data networks have a higher mean Internet speed and less variability (assuming that newer mobile devices and 4G LTE networks have more reliable and consistent data speeds) than 2G mobile devices.

In place of the interactive demonstration described above, you can present examples of static images taken from the figures in this chapter, or from other sources. Instead of dynamically generating distributions to represent the intended scenarios, students can be asked to interpret what the given figures may represent or how they

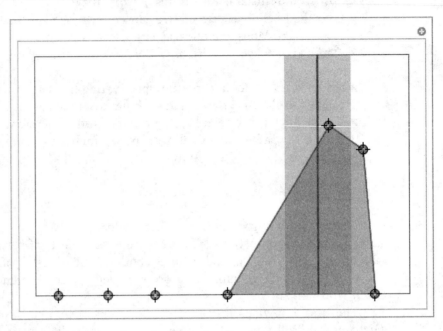

Figure 4.1. Mean and standard deviation of a skewed distribution.

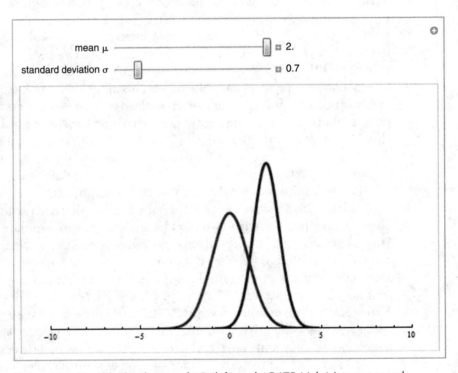

Figure 4.2. Possible distribution of 2G (left) and 4G LTE (right) Internet speeds.

reflect underlying differences in the data. However, if you desire to capture the dynamic nature of this visualization without the use of technology, you can serve as the "computer-generated image" through the use of a piece of chalk and the chalkboard. Have students define populations as you draw the distributions on the board. Guiding students with questions to get them to alter the variability of a population can help them to see changes and better visualize this concept. After a few examples, students can then be asked to generate distributions at their desks.

ASSESSMENT Multiple-choice or short-answer quizzes with the following conceptual questions can help determine how well students understand variance and *SD*.

- What is a deviation?
- What does a positive/negative deviation tell us?
- What does a small/large deviation tell us?
- What is the sum of squares, and why do we need it?
- Why do we have to square the deviation?
- How is the *SD* related to the variance?
- How does the shape of a distribution of scores affect the *SD*?
- From a small set of data, calculate the variance and *SD*.

DISCUSSION Students may be surprised when their instructor asks them to get out their mobile devices! This activity can generate a lot of interest in students who are curious (and sometimes competitive) about their Internet speeds. A number of students may not know their Internet-provider's advertised Internet speeds, and most students have never tested their actual real-time Internet speeds, especially in the classroom. Other data from mobile devices, such as screen size, number of installed apps, number of photos stored on the device, social media posts in the past 24 hours, and so on, also can be collected for calculations of variability.

Because classroom technology can be unpredictable at times, you should make sure that you are comfortable with the technology before planning to use it in class, for example, by practicing with the demonstrations from the Wolfram Demonstration Project on the classroom computer. Some students may also need additional help downloading the Internet connection speed app to their mobile device.

REFERENCES McClelland, G. H. (2015). *The normal distribution*. Retrieved from the Wolfram Demonstrations Project website: http://demonstrations.wolfram.com/TheNormalDistribution/

Smith, A. (2015, April). *U.S. smartphone use in 2015*. Retrieved from the Pew Research Center website: http://www.pewinternet.org/2015/04/01/us-smartphone-use-in-2015/

Wolfram, S. (2015). *Mean and standard deviation of a distribution*. Retrieved from the Wolfram Demonstrations Project website: http://demonstrations.wolfram.com/MeanAndStandardDeviationOfADistribution/

5

GETTING DICEY: THINKING ABOUT NORMAL DISTRIBUTIONS AND DESCRIPTIVE STATISTICS

Robert McEntarffer and Maria Vita

In the activity described in this chapter, students learn about how frequency distributions are organized, why some distributions tend to be normal, and how measures of central tendency and variability apply to distributions created by the students.

CONCEPT
Many psychology students are intimidated by the statistics used in psychology research. A basic understanding of the normal curve and a conceptual understanding of variance and statistical significance are important for students who will be interpreting research results. During the activity we present in this chapter, students roll dice and use the resulting data to construct their own distributions with data they gather. This hands-on experience can help demystify how data are represented in frequency distributions and normal curves as well as reduce the intimidation factor for students who may otherwise wish to avoid the quantitative elements of research methodology.

MATERIALS NEEDED
Each pair of students needs two dice. Students also need paper and markers (or graphing software) to draw their distributions. Alternatively, you can use online resources to collect and graph the class data. In this case, a group spreadsheet or online survey prepared before the activity would allow you to present data from the entire class and graphically represent these data.

INSTRUCTIONS
This activity can be used before or after a discussion of descriptive or inferential statistics to strengthen students' understanding of basic statistical concepts. First, organize students into pairs, giving each pair of students two standard dice. Tell the pairs that the person with the dice is the "roller" and the other person is the "recorder." Explain to the pairs that their goal is to roll the dice, add the numbers on the two dice, and record the sum. They should do this as many times as they can in a 5-minute period (shorter or longer if you prefer). The length of time you give the pairs to roll the dice will determine their sample size and how normal their distribution will be. Make sure that each team rolls at least 30 times. After the allotted time has expired, have the teams organize their data into a frequency distribution (see Table 5.1 for an example).

http://dx.doi.org/10.1037/0000024-006

Activities for Teaching Statistics and Research Methods: A Guide for Psychology Instructors, J. R. Stowell and W. E. Addison (Editors)

Table 5.1 *Frequency Distribution Example*

Score	Frequency
2	3
3	4
4	4
5	6
6	6
7	9
8	8
9	5
10	4
11	4
12	2

Have the student pairs graph their frequency distributions by hand or in Excel; they can make histograms or line graphs. After collecting all the graphs, show some or all of them to the class (e.g., using a document camera or posting them around the classroom). All of the graphs should approximate a normal curve, although the distributions from the pairs with the larger sample sizes (the most rolls) should more closely approximate a normal distribution. Students can discuss why the distributions look different, which will lead to a discussion of how sample size influences the distributions and what tends to happen as sample size increases (the distributions become more normal). You can ask students why the scores are distributed in this way (scores of 7 and 8 are more likely than scores of 2 and 12) and discuss examples of other data that are likely to be distributed normally, such as people's heights, IQ scores, and scores on a measure of anxiety.

Next, draw a curve on the board that represents the theoretical results from many dice rolls (or project an image of this kind of curve). This curve will be fairly normal (bell shaped) because of the probabilities of getting different scores from rolling and adding two dice. Alternatively, you can use one of the graphs the students created, as long as it represents a fairly normal distribution. Discuss with students the following statistical concepts using the normal curve:

- *Measures of central tendency (mean, median, mode).* Students should realize that the mode (the most frequent score) is obvious from the frequency distribution and their graph. The class could discuss how to use the distribution to find the median and mean. Pairs of students could compute the median and mean, if desired, but the focus of this activity is more conceptual than computational. Students should be able to recognize that the mean, median, and mode are all equal in a perfectly normal curve.
- *Measures of variability (range, variance, standard deviation).* Students should be able to recognize the range immediately ($12 - 2 = 10$). The class could discuss how the curve relates to standard deviation by posing such questions as "What would the curve look like if the standard deviation were very small?" (narrow) and "What would the curve look like if the standard deviation were very large?" (wide).

The hands-on nature of this activity may help some statistics-averse students overcome any anxiety or discomfort associated with analyzing quantitative data. Gathering, graphing, and thinking about their own data may help make these statistical concepts more real for students. After this activity, students should have a more complete understanding of how to organize data into a frequency distribution and a histogram, how sample size influences a distribution, and how to interpret basic descriptive statistics (e.g., mean, median, mode, as well as variance/standard deviation) from a distribution.

In addition, this activity can be used to help students think about inferential statistics. You could draw a new distribution in a contrasting color on the curve used for the class discussion. The new distribution should be skewed, starting toward the middle and bottom of the graph and ending at the top right (indicating that the mode is 11 or 12; see Figure 5.1). Pose the following question to the class: "What would you conclude if a pair of students got a graph that looks like this, with a mean of about 10? Keep in mind that the mean for rolling a pair of dice in this activity is typically about 7." Students can begin the discussion in pairs, or you can have a discussion that includes the whole class. Students will talk about how unlikely this new distribution is (e.g., "That's not possible," "They must be cheating," "There must be something wrong with the dice"). At this point you can discuss the role of probability in generating a normal distribution and why a skewed distribution is unlikely to occur when the dice rolling is truly random. This discussion can lead into a conceptual discussion about statistical significance; for example, the results of a two-sample study are called statistically significant when the

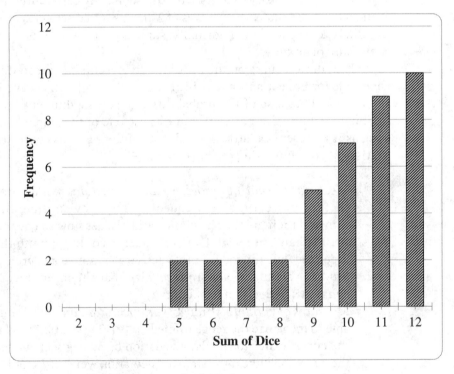

Figure 5.1. A skewed frequency distribution.

two distributions of scores are so different that it is very unlikely that the difference could have occurred by chance.

In a variation of this activity, you could use a Google form (see a YouTube demo at https://youtu.be/EA_KbtsX3Kk and step-by-step instructions at http://tinyurl.com/diceinstructions) or Microsoft Excel to collect frequency distribution data from each pair of students. You can use these data to illustrate digitally a class histogram that includes hundreds of rolls. This demonstration provides further evidence of how sample size influences the shape of a distribution. The results from any pair of students can be compared to the distribution for the whole class, demonstrating that the larger the sample size, the more normal the distribution is likely to be (e.g., the class distribution [gray bars in Figure 5.2] compared with one pair of students [black bars in Figure 5.3]).

This activity emphasizes a conceptual understanding of basic statistical concepts rather than requiring students to compute any of the statistics, although the activity could be modified to include computation as well. It may be useful for students to understand why concepts such as frequency distributions, measures of central tendency, and variability are important before they learn to compute them. Knowing the "why" of these basic statistics may help motivate students to persevere as they learn how to use statistics later in their studies, and this conceptual understanding should help students think more critically about research results they encounter in other psychology courses or in popular media sources.

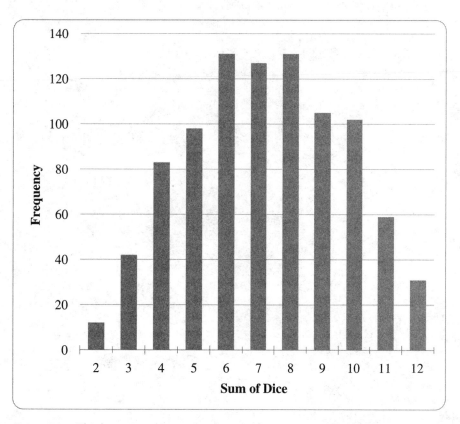

Figure 5.2.   *This large sample size closely approximates a normal distribution.*

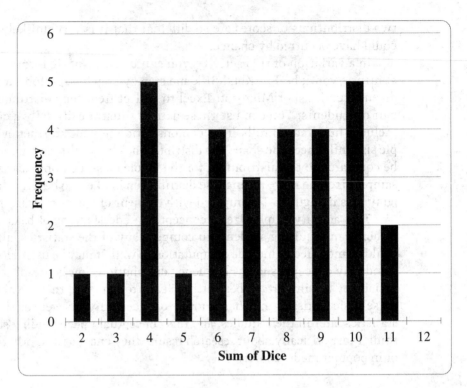

Figure 5.3. A small sample size is less likely to approach a normal distribution.

A Low-Anxiety Introduction to the Standard Normal Distribution and Measures of Relative Standing

Laura Brandt and William E. Addison

The activity described in this chapter was designed as a nonthreatening, step-by-step approach to introducing students to the standard normal distribution and the use of percentiles and to have them apply these concepts to data derived from everyday experiences. Oftentimes students memorize the basic characteristics of the standard normal distribution but fail to understand how the distribution applies to situations they are likely to experience in their own lives. A better understanding of the standard normal distribution can help students interpret real-life data and apply related concepts, such as probability, to inferential statistics.

CONCEPT

In a typical statistics course students' initial exposure to z scores and percentiles occurs in the context of measures of relative standing. Once students have a basic understanding of these measures, the standard normal distribution can be explained as a symmetric, unimodal distribution of z scores, with identifiable areas under the curve. To be specific, 68.3% of the scores fall within 1 standard deviation (*SD*) of the mean, 95.4% fall within 2 *SD*s, and 99.7% fall within 3 *SD*s.

Once students understand the basic features of the normal distribution, they can apply that knowledge to real-life data such as IQ scores or ACT/SAT scores. For example, assuming that IQ scores are normally distributed in the population, with a mean of 100 and an *SD* of 15 (these figures may vary slightly depending on the particular test), an IQ score of 115 falls 1 *SD* above the mean. Because z scores are in *SD* units, an IQ score of 115 corresponds to a z score of 1.00, or 1 *SD* above the mean, which is equivalent to a z score of 0.00. Also, approximately 34% (half of 68%) of the distribution falls between 115 and the mean of 100 (see Figure 6.1). By adding the lower 50% of the distribution students can see that the percentile rank (i.e., the percentage or proportion below a particular score) for a score of 115 is equal to .84; in other words, an IQ score of 115 falls at the 84th percentile, or a percentile rank of .84.

The same percentile rank (.84) applies to any score from a normal distribution that falls 1 *SD* above the mean. For example, if ACT scores are normally distributed in the population, with a mean of 20 and an *SD* of 5, an ACT score of 25 ($z = 1.00$) also corresponds to a percentile rank of 84. Similarly, an ACT score of 15 ($z = –1.00$) would correspond to a percentile rank of 16, because approximately 34% of the distribution falls between 1 *SD* below the mean and the mean of 20 ($z = 0.00$). These kinds of simple examples should help alleviate some of the anxiety students experience when they are

http://dx.doi.org/10.1037/0000024-007
Activities for Teaching Statistics and Research Methods: A Guide for Psychology Instructors, J. R. Stowell and W. E. Addison (Editors)

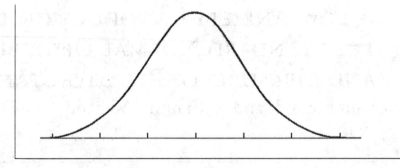

IQ score	55	70	85	100	115	130	145
z score	−3.00	−2.00	−1.00	0.00	1.00	2.00	3.00
Percentile rank	.001	.023	.16	.50	.84	.977	.999

Figure 6.1. IQ scores and the normal distribution.

initially exposed to the application of statistics to behavioral data, whether it is in a basic statistics course or an introductory psychology class.

MATERIALS NEEDED AND INSTRUCTIONS

In class, provide students with the worksheet found in Appendix 6.1 (without the answers shown in parentheses) and ask them to complete the figure and respond to the five questions. This first worksheet is designed to give students the opportunity to assess their understanding of the basic characteristics of the normal distribution. After giving students enough time to complete the worksheet (10–15 minutes is typical), collect the worksheets and go over the answers in class. You should remind students that if they know the distribution of scores is normal, and they know the mean and standard deviation, they can apply their knowledge of the normal distribution to that data set. The worksheet presented in Appendix 6.2, which incorporates real-life data in the form of SAT scores, is slightly more challenging than the first worksheet and may be best used as a homework assignment. The worksheet presented in Appendix 6.3, which requires students to compare scores on several different distributions, is particularly appropriate for a group activity in which small groups (three students is ideal) work together to answer the questions. The three worksheets are designed to be increasingly difficult so that you can introduce students to the more challenging questions in a gradual manner, in an effort to minimize the anxiety that some students experience with exercises of this kind. In addition, students have help from their group members in answering the hardest questions, which are found in the third worksheet (see Appendix 6.3).

ASSESSMENT

The most straightforward measure of student learning in this case would be a pretest–posttest measure of students' knowledge of z scores, percentiles, and the standard normal distribution. Most basic statistics textbooks have relevant exercises from which five to six appropriate questions could be drawn to create such a measure. Anecdotal evidence from our own students suggests that the more seriously they take these assignments/activities, the more likely they are to learn the basic concepts.

Students often struggle to make the connection between the standard normal distribution and real-life data that may have an impact on their own lives, such as physical (e.g., height, blood pressure) and behavioral (e.g., IQ, test scores) traits that are assumed to be normally distributed in the population. By using examples that incorporate scores from these types of distributions you can emphasize the real-life applications of statistics, a strategy designed to increase students' interest in the material.

Because proportions under the standard normal curve are *theoretical* proportions (i.e., based on an infinite number of cases), they are literally the same as probabilities. Thus, examples that incorporate proportions under the standard normal curve can provide an easy transition into a discussion of probability as it applies to the use of inferential statistical tests. For example, asking the question "What proportion of SAT Math scores fall above 700?" is the same as asking, "What is the probability that a randomly selected SAT Math score will fall above 700?"

Advanced versions of this activity may include having students use the formula for converting a raw score into a z score, which can provide the basis for more challenging problems that incorporate scores from actual or hypothetical psychological tests. The presentation found on the Math Is Fun website (see the Resources section, below) can be a good starting point for these kinds of problems.

- A straightforward and low-key presentation that describes the normal distribution can be found on the Math Is Fun website (see http://www.mathsisfun.com/data/standard-normal-distribution.html). The discussion includes real-life examples of the normal distribution (e.g., heights of people, blood pressure) as well as the formula for converting raw scores to z scores. Teachers can use the information here to supplement their own presentations.
- A more advanced presentation of the normal distribution can be found on the StatTrek website (see http://stattrek.com). This site provides a description of the normal distribution as a family of continuous probability distributions and includes the equation for defining the distribution.

Appendix 6.1

A Normal Distribution of Anxiety Scores

Given that scores on a test designed to measure anxiety level are normally distributed in the population, with a mean of 60 and a standard deviation of 5, you are to complete the table below by providing the following information:

- the anxiety scores,
- the z scores from –3.00 to 3.00, and
- the percentile rank for each z score.

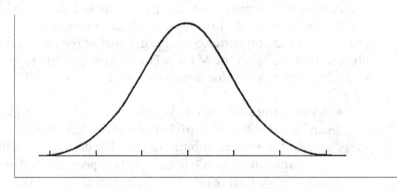

Anxiety score	(45)	(50)	(55)	(60)	(65)	(70)	(75)
z score	(–3.00)	(–2.00)	(–1.00)	(0.00)	(1.00)	(2.00)	(3.00)
Percentile rank	(.001)	(.023)	(.16)	(.50)	(.84)	(.977)	(.999)

Based on the completed table, you are to answer the following questions:

- What is the z score that corresponds to an anxiety score of 55? (–1.00)
- What is the z score that corresponds to an anxiety score of 70? (2.00)
- An anxiety score of 65 corresponds to what percentile rank? (P.84, or the 84th percentile)
- An anxiety score of 50 corresponds to what percentile rank? (.023, or about the 2nd percentile)
- Given that an anxiety score exceeding 2 standard deviations above the mean places a person at risk for diagnosis of an anxiety disorder, what proportion of the population would be considered at risk for an anxiety disorder? (.023, or 2.3%, which is 1 – .977).

Appendix 6.2

SAT Scores and the Normal Distribution

Given that scores on the SAT Math (SAT M) test are normally distributed in the population, with a mean of 500 and a standard deviation of 100, you are to complete the table below by providing the following information:

- the SAT M scores,
- the z scores from –3.00 to 3.00, and
- the percentile rank for each z score.

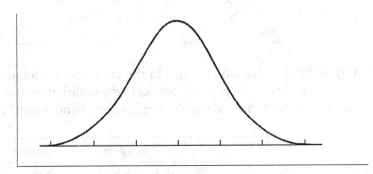

SAT M score	(200)	(300)	(400)	(500)	(600)	(700)	(800)
z score	(–3.00)	(–2.00)	(–1.00)	(0.00)	(1.00)	(2.00)	(3.00)
Percentile rank	(.001)	(.023)	(.16)	(.50)	(.84)	(.977)	(.999)

Based on the completed table, you are to answer the following questions:

- What is the z score that corresponds to an SAT M score of 300? (–2.00)
- An SAT M score of 400 corresponds to what percentile rank? (P.16, or the 16th percentile)
- What proportion of the distribution falls between SAT M scores of 400 and 600? (The proportion between SAT M scores of 400 and 600 is equal to .68, because this is the area of the standard normal distribution within 1 standard deviation of the mean.)
- What is the median of this distribution of SAT M scores? (The median of the distribution is 500. In any distribution [normal or not] the median corresponds to the 50th percentile, which is 500 in this case. Additionally, in a normal distribution, the three measures of central tendency [mean, median, and mode] are equal; in this case, the three measures are equal to 500.)
- Applicants to State University must score above the 90th percentile to be considered for the University Honors Program. What proportion of the population would meet this criterion? (The proportion of the population that would meet this criterion is .10, or 10%. The 90th percentile means that .90 or 90% of the population scores at or below this point, so the rest of the distribution [1 – .90] will score above this point.)

Appendix 6.3

Comparing Scores on Midterm Exams

You and two of your friends have just taken your midterm exams, and each of you received a score of 50 on your respective exam. You begin to argue over which of you did "the best" on his/her exam. Your instructors have indicated that the scores for all three exams are normally distributed.

Exam	Mean	Standard deviation	Your scores
Psychology	60	5	50
Chemistry	30	10	50
Algebra	52	2	50

Given the information provided in the table above, and using the percentile ranks in Figure 6.1, you are to fill in the blanks in the table below and explain how the scores can be interpreted. Which of you did the best on his/her exam? How can you tell?

	Psychology	Chemistry	Algebra
z score	(−2.00)	(2.00)	(1.00)
Percentile rank	(.023)	(.977)	(.16)

(The student who took the chemistry exam did the best, scoring better than 97.7% of students who took the exam, with a score that was 2 standard deviations above the mean [z = 2.00]. The student who took the Algebra exam was 1 standard deviation below the mean [z = −1.00], better than 16% of other students taking the exam. The psychology student scored 2 standard deviations below the mean [z = −2.00], better than just 2.3% of students who took the exam.)

USING THE HEAT HYPOTHESIS TO EXPLORE THE STATISTICAL METHODS OF CORRELATION AND REGRESSION

George Y. Bizer

In the activity described in this chapter, students enter data into a computer, create scatterplots, compute Pearson product–moment correlation coefficients, and conduct linear regression analyses. Follow-up activities may include adding or removing outliers or creating new data sets for further analysis.

CONCEPT

There are three primary routes through which to explore data that contain two or more continuous variables: (a) a *scatterplot*, which demonstrates visually the association between two variables; (b) the Pearson product–moment correlation coefficient (hereafter *Pearson correlation coefficient*), which quantifies the association; and (c) a *regression analysis*, which allows one to predict a specific value of one variable given a value of the other variable. In the activity presented in this chapter, students gain practice with all three techniques by exploring data that demonstrate a correlation between heat and crime, a well-documented phenomenon (the *heat hypothesis*, e.g., Anderson, 2001; Ranson, 2014).

MATERIALS NEEDED AND INSTRUCTIONS

Students need access to a computer with Microsoft Excel or SPSS. Students enter the data from Table 7.1 into Excel or SPSS. The data from van Zanten (2013) represent the high temperature (°F) and total number of crimes in Chicago, Illinois, on the first day of each month during 2010. Using the instructions given below for Excel or SPSS, students create a scatterplot, calculate the Pearson correlation coefficient, and conduct a linear regression analysis to examine the association between the temperature and crime figures.

Using Excel 2010

To create the scatterplot, students click the upper left data cell in Table 7.1 (January's high temperature, 16°F) and then drag to the lower right data cell (910), creating a highlighted 2 × 12 grid. Then, under the "Insert" tab, students select the type of chart titled "Scatter With Only Markers." A scatterplot should then appear. Students will note that the data points generally fall in a pattern from the lower left to the upper right region of the chart, suggesting a positive correlation.

---

http://dx.doi.org/10.1037/0000024-008
Activities for Teaching Statistics and Research Methods: A Guide for Psychology Instructors, J. R. Stowell and W. E. Addison (Editors)

Table 7.1 *Data on Month, Temperature, and Number of Crimes Committed*

Month	Temperature (°F)	Number of crimes
January	16	1,178
February	30	979
March	36	1,085
April	83	1,212
May	70	1,149
June	85	1,216
July	77	1,202
August	86	1,139
September	81	1,178
October	73	1,198
November	54	1,077
December	26	901

Note. Table values are based on data from van Zanten (2013).

To calculate the Pearson correlation coefficient, students type *=correl(* into an empty cell anywhere on the Excel worksheet. Students then select all the cells with data in the "Temperature" column for array1 and all the cells with data in the "Number of Crimes" column for array2. The Pearson correlation coefficient, .67, should appear in this cell.

To conduct the regression analysis, the Excel Analysis ToolPak software must be installed on each student's laptop or tablet (see, e.g., http://www.excel-easy.com/data-analysis/analysis-toolpak.html). Once this has been done, students click the "Data Analysis" button under the "Data" tab and then select "Regression" from the dialogue box. To input the Y range, they select the cells in the "Number of Crimes" column, and to input the X range, they select the cells in the "Temperature" column. A new sheet will appear with regression values. Students will note that the multiple R value of .67 replicates the value found in the prior analysis, that the significant F value of .02 (rounded from .01629) indicates that the association between temperature and crime is statistically significant, and that the "Coefficients" for "X Variable 1" of 2.58 indicates that each degree increase in temperature is associated with 2.58 additional crimes committed.

Using SPSS

To create the scatterplot, students select "Graphs" from the main menu and then select "Chart Builder." From there, students select "Scatter/Dot" from the "Gallery" options. They then click the upper left graph option ("Simple Scatter") and drag it to the "Drag a Gallery Chart Here" field. They then drag "Temperature" to the *x*-axis field and "Number of Crimes" to the *y*-axis field. Finally, students click "OK." A scatterplot should then appear in the output file. Students will note that the data points generally fall in a pattern from the lower left to the upper right region of the chart, suggesting a positive correlation.

To calculate the Pearson correlation coefficient, students select "Analyze" from the main menu, then "Correlate," then "Bivariate." They then drag both "Temperature" and "Number of Crimes" into the "Variables" field. They then click "OK." The Pearson correlation coefficient, .674, should appear in the output file.

To conduct the regression analysis, students select "Analyze" from the main menu, then "Regression," then "Linear." They should drag "Temperature" into the "Independent(s)" field and "Number of Crimes" into the "Dependent" field and then click "OK." Students will note in the output file that the R value of .67 (found in the second table) replicates the value found in the correlation analysis, the "sig" value of .016 (found in the third table) indicates that the association between temperature and crime is statistically significant, and the "Unstandardized Coefficient B" for "Temperature" of 2.58 (found in the fourth table) indicates that each degree increase in temperature is associated with 2.58 additional crimes committed.

Important Concepts for Students

Students should understand that these three activities all explain the association between crime and temperature, but they do so with different levels of precision. The scatterplot provides the most basic information in that the values merely *appear* to be positively correlated. The Pearson correlation coefficient ($r = .67$) quantifies the relationship. The probability value ($p = .02$) confirms that the association is statistically significant. The regression analysis provides additional information, indicating that each increased degree of temperature is associated with 2.58 additional crimes committed per day.

ASSESSMENT After completing this activity, students should be able to create and interpret scatterplots, Pearson correlation coefficients, and regression tables. To assess student learning, you could create an additional set of data and ask students to create and interpret a scatterplot, the Pearson correlation coefficient, and the regression analysis. Alternatively, you could simply ask students to explain and interpret already-completed analyses.

DISCUSSION Outliers can have a profound influence on the association between two variables, particularly in a small data set like this one. Students can be asked to identify which one of the 12 data points appears to be an outlier by examining the scatterplot and/or the raw data. Note in Table 7.1 that January 1 appears to be an outlier: It was a cold day, but there was a relatively large number of crimes committed. You can ask students why this day may have been an exception to the otherwise strong association. Then ask them to recalculate the Pearson correlation coefficient and regression analyses without including this data point. Doing so increases the coefficient from .67 to .90! You may then invite students to reinclude the January data point and add additional, fictitious data points one at a time in attempts to increase or decrease the association between the two variables.

Once students have learned from the temperature/crime data file, they can create and explore additional data files. You can ask them to find their own files from online resources, or prepare and conduct a survey through which they obtain data from friends or classmates.

Scatterplots, Pearson correlation coefficients, and regression analyses are appropriate when testing the association between two *continuous* variables. It is inappropriate to use such analyses when the variables are *discrete* (categorical) in nature. For example, consider a study in which people are asked to indicate their sex (male or female) and preferred ice cream flavor (chocolate, vanilla, or strawberry). Because both variables are discrete, assessing the relation between the two requires a chi-square analysis rather than a Pearson correlation coefficient or regression analysis. Therefore, data files for the activity described here should include continuous variables only. In addition, students should be told that proper data analysis techniques are determined not by the type of research (correlational or experimental) but instead by the type of variables examined in that research (discrete or continuous; see Fidell & Tabachnick, 2003, for more information on the difference between discrete and continuous variables).

The correlational nature of the data used in this activity also means that no causal conclusions can be drawn. Is it conceivable that the results emerge because heat increases aggression? Yes, and indeed much research has argued for a causal association. However, it also is possible that hours of daylight (which is positively correlated with temperature) might be responsible for the association found in these data. A variety of other factors may also play a causal role. Thus, students should be reminded that correlation does not connote causation, regardless of how clean the scatterplot may appear or how small the probability value is.

REFERENCES

Anderson, C. A. (2001). Heat and violence. *Current Directions in Psychological Science,* *10,* 33–38. http://dx.doi.org/10.1111/1467-8721.00109

Fidell, L. S., & Tabachnick, B. G. (2003). Preparatory data analysis. In J. A. Schinka, W. F. Velicer, J. A. Schinka, & W. F. Velicer (Eds.), *Handbook of psychology: Vol. 2. Research methods in psychology* (pp. 115–141). Hoboken, NJ: Wiley.

Ranson, M. (2014). Crime, weather, and climate change. *Journal of Environmental Economics and Management, 67,* 274–302. http://dx.doi.org/10.1016/j.jeem.2013.11.008

van Zanten, E. (2013). *Chicago crime vs. weather* [Data set]. Retrieved from http://crime.static-eric.com/data/weather/2010.csv

ACTIVE LEARNING FOR UNDERSTANDING SAMPLING DISTRIBUTIONS

David S. Kreiner

The concept of a sampling distribution can be difficult to understand, yet it is crucial for learning inferential statistics. In this chapter a class activity is described in which students generate sample means by selecting scores from a population. The activity can help students understand the logic of drawing conclusions about a hypothetical distribution.

CONCEPT

A *sampling distribution* is a hypothetical frequency distribution composed of sample statistics from an infinite number of samples drawn from a population. The sampling distribution is a crucial concept for students learning about research methods and statistics (Kett, 2011). The hypothetical nature of the distribution makes it challenging for students to understand its characteristics (Mulekar & Siegel, 2009). For example, it may be difficult for students to understand why one would imagine drawing repeated samples from a population when researchers usually measure only one sample (Turner & Dabney, 2015).

The simulation activity described here provides direct experience in drawing samples and observing the distribution of sample statistics. The goal is for students to better understand the concept of a sampling distribution and why it is logical to make certain assumptions about its mean, standard deviation, and shape.

MATERIALS NEEDED

The activity requires a set of 500 slips of paper, each with a printed number. The numbers represent individual scores in a population. I recommend generating the set of numbers in advance of the activity and reusing the numbers when the activity is performed in future classes. On the basis of the minimum and maximum values of the variable (e.g., 1–10), you should prompt the class to suggest a variable that the numbers will reflect (e.g., scores on a quiz). Identifying a specific variable will make the activity more meaningful.

The population scores can be generated using the "randbetween" function in Excel, which allows specification of the minimum and maximum values of the variable. The function is copied to 500 cells, which creates a list of 500 random numbers. The numbers should be printed in a large font and cut so that each number appears on a separate slip of paper. Place the slips of paper into a container that is suitable for passing around the classroom, such as a tea tin or small bag.

http://dx.doi.org/10.1037/0000024-009
Activities for Teaching Statistics and Research Methods: A Guide for Psychology Instructors, J. R. Stowell and W. E. Addison (Editors)

Begin by defining the term *sampling distribution* and pointing out that it is a difficult concept because it is hypothetical. It is impossible to generate an infinite number of samples, but we can imagine that process. Emphasize the importance of understanding the concept of the sampling distribution. To fully understand statistical significance it is necessary to have a grasp of how probability theory applies to sample statistics. The sampling distribution is the key concept linking samples to populations. Therefore, students will have difficulty learning about significance testing if they do not first develop an understanding of sampling distributions.

You should then present the container of scores and explain that each slip of paper represents the score of one research participant on a variable of interest. The entire set of numbers in the container represents a *population*. Explain to the students that if a researcher selects a few slips of paper, those numbers represent a *sample*. It is helpful to remind students that the same numbers can be repeated on different slips of paper, just as multiple individuals in a population could achieve the same score on a variable. Furthermore, students should not attempt to edit their samples by discarding repeated numbers or sequences that appear to be nonrandom. This is a good opportunity to remind students that random sequences of numbers can include repeated scores.

The students take turns drawing samples of a specified size out of the container. A sample size of 10 simplifies the process of calculating sample means. After the first student has drawn 10 scores, he or she should read the numbers aloud while you write them on the board. The rest of the class calculates the sample mean, which you should record in a separate location on the board. The student returns the scores to the container and mixes up the slips of paper to randomize them.

With one sample mean recorded, ask your students the following two questions:

1. *What is your best guess about the mean of the entire population?* Students should indicate that the sample mean is the best estimate of the population mean.
2. *Would you be surprised to find out that the population mean is not exactly the same as the mean of this one sample?* Students should respond that they would not be surprised, because the sample did not contain the entire population.

Another student then draws a second sample from the container. Record the second mean beneath the first sample mean and note that, having drawn two samples from the same population, the class has begun to create a sampling distribution. Then ask the following two questions:

1. *What is your best guess now about the mean of the population?* Students should arrive at the idea of averaging the two sample means.
2. *Why do the sample means differ from each other?* Students should note that the scores that end up in each sample are determined randomly. The variability of sample means is called *sampling error* and is key to understanding how probability is used in inferential statistics.

The activity continues, with each student drawing a sample, calculating the mean, and adding it to the list of means. As the means accrue, students will notice that most are close to each other, with occasional means that are much lower or higher. After all of the students have completed their samples, ask them the following questions:

1. *What is your best guess now about the mean of the population? How confident are you that it is close to the actual value? If we could have drawn an infinite number of samples, what could you logically conclude?* Students should recognize that the mean of the sample means is a good estimate of the population mean and becomes a better estimate as the number of samples increases. With an infinite number of samples, the mean of the sampling distribution will equal the mean of the population.

2. *If we want to measure how spread out the sample means are from each other, what statistic could we calculate?* When students suggest calculating the standard deviation, point out that it has a special name when calculated for a sampling distribution: the *standard error*. If time allows, have the class calculate the standard deviation of the sampling distribution and then compare it to the standard deviation of the sample divided by the square root of N. The two numbers should be similar, especially with a large number of samples.

3. *If we graphed the frequency distribution of sample means, what would you expect it to look like?* An introductory statistics class will generally have already learned about normal distributions, and thus students should suggest that the distribution will be approximately normal. You can point out that the distribution will become more normal as the sample size increases (the *central limit theorem*). For example, drawing samples of 30 scores will yield a more normally shaped sampling distribution than samples of 10. Also, as the number of samples increases, the sampling distribution will become more normal.

ASSESSMENT

To assess learning, ask students to respond to short-answer questions in the form of a laboratory exercise, quiz, or homework assignment. Consider including both comprehension questions and application items that prod students to think further about sampling distributions (see Appendix 8.1). Students should also be able to perform better on questions later in the course about significance testing and probability (e.g., "Explain how the sampling distribution is used to construct a confidence interval for the mean").

DISCUSSION

The literature on teaching sampling distributions includes a variety of simulation activities, some with evidence of student learning (Aguinis & Branstetter, 2007; Dyck & Gee, 1998). The activity described here is unique in that it requires each student to take a sample of numerical data in the context of a meaningful research scenario. Students should find the activity more engaging than simply listening to a lecture about sampling distributions.

With a large class, the amount of time necessary for all students to participate may be impractical. One option is for students to split into small groups and complete the activity. Another potential problem is lack of involvement on the part of students who are not currently taking a sample. A solution is to involve students in other roles, such as writing scores on the board or calculating the means. The distribution of sample means will be more realistic if students take care to sample randomly from the container. Small slips of paper will sometimes stick to each other or fall on the floor, so it helps to have another student assisting.

You may wish to illustrate sampling from two populations by generating two sets of numbers, as in an experimental design with a treatment group and a control group. Each student draws a sample from each container and calculates the difference between means. You may even create a sense of mystery by commenting that the two containers may or may not have the same mean. Just as in real research, a researcher can arrive at a conclusion about whether the population means are different by comparing the difference between means in one research project to a hypothetical sampling distribution of differences between means.

Although the sampling distribution is a hypothetical concept, hands-on experience in simulating the distribution can help students appreciate the differences among sampling distributions, sample distributions, and population distributions, with each type of distribution playing a unique role in significance testing. You can refer back to the activity later in the course as students learn about confidence intervals and significance testing.

REFERENCES

Aguinis, H., & Branstetter, S. A. (2007). Teaching the concept of the sampling distribution of the mean. *Journal of Management Education, 31*, 467–483. http://dx.doi.org/10.1177/1052562906290211

Dyck, J. L., & Gee, N. R. (1998). A sweet way to teach students about the sampling distribution of the mean. *Teaching of Psychology, 25*, 192–195. http://dx.doi.org/10.1207/s15328023top2503_6

Kett, J. R. (2011). Teaching sampling distributions using Autograph. *The Mathematics Teacher, 105*, 226–229. http://dx.doi.org/10.5951/mathteacher.105.3.0226

Mulekar, M. S., & Siegel, M. H. (2009). How sample size affects a sampling distribution. *The Mathematics Teacher, 103*, 34–42.

Turner, S., & Dabney, A. R. (2015). A story-based simulation for teaching sampling distributions. *Teaching Statistics, 37*, 23–25. http://dx.doi.org/10.1111/test.12067

Appendix 8.1

Sample Assessment Questions

COMPREHENSION QUESTIONS

1. Explain how a sampling distribution could be constructed.
2. How is a sampling distribution different from a sample distribution?
3. How is a sampling distribution different from a population distribution?
4. What can we logically conclude about the mean of a sampling distribution?
5. What causes the means to differ from each other in a sampling distribution?
6. What can we logically conclude about the shape of a sampling distribution?
7. What is the name for the standard deviation of a sampling distribution?

APPLICATION QUESTIONS

1. What would you expect the sampling distribution to look like if you took very small samples (e.g., $N = 3$) from the population? Why?
2. If the sampling distribution is normal, what can we conclude about the percentage of sample means that would be more than 2 standard deviations above or below the mean?
3. If you increase your sample size, what will happen to the size of the standard error? Why?
4. Describe how you would construct a sampling distribution for a sample statistic other than the mean (e.g., for the median).
5. A classmate claims that, because a sampling distribution is imaginary, there is no way to know what its mean would be. How would you refute that claim?

9 | TESTING STUDENTS FOR ESP: DEMONSTRATING THE ROLE OF PROBABILITY IN HYPOTHESIS TESTING

William E. Addison

Students in introductory statistics classes often have difficulty understanding how probability is used to make a decision about the hypotheses in a psychology study. Using an activity that tests students' ESP ability, instructors can reinforce the notion that conclusions about the alternative or research hypothesis are based on the probability that a finding consistent with the null hypothesis could have occurred by chance.

CONCEPT

Although null hypothesis significance testing has its critics (see, e.g., Cohen, 1994; Nickerson, 2000; Trafimow & Marks, 2015), it remains a critical element in students' understanding of the conduct and reporting of psychological research. A key concept in understanding null hypothesis significance testing is the notion of *indirect confirmation*, in which conclusions about the alternative (or research) hypothesis are based on the probability that the null hypothesis is true. For example, a difference between two sample means is considered "significant" when the probability is very small that the difference occurred by chance (i.e., the null hypothesis is true), generally less than or equal to an alpha level of .05 or .01. The activity described in this chapter is designed to introduce students to this concept by testing them for one of several forms of ESP. In addition, the activity provides a review of z scores and the standard normal distribution.

MATERIALS NEEDED

The only materials needed for this activity are a deck of Zener cards, designed in the early 1930s by Karl Zener (1903–1963). Zener was an associate of Joseph Rhine (1895–1980), who studied parapsychology at Duke University in the 1930s and 1940s and who coined the term *extrasensory perception* (Rhine, 1934). A standard deck of Zener cards contains 25 cards, five cards picturing each of the five symbols shown in Figure 9.1.

The cards can be purchased, generally for less than $10, online or at a magic store, or they can be made by hand or with the aid of a computer graphics program. Before beginning the demonstration you should show the students what the five symbols look like so they know the symbols they will be trying to perceive. I present PowerPoint slides that provide a brief history of the study of ESP, definitions of the three forms of ESP (telepathy, clairvoyance, and precognition), and pictures of the five symbols on the cards.

http://dx.doi.org/10.1037/0000024-010
Activities for Teaching Statistics and Research Methods: A Guide for Psychology Instructors, J. R. Stowell and W. E. Addison (Editors)

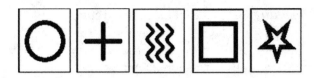

Figure 9.1. Five symbols used in a deck of Zener cards. From *Vector Drawing of the Five Kinds of Zener Cards*. Wikimedia Commons. Retrieved from https://commons. wikimedia.org/wiki/File:Cartas_Zener.svg. In the public domain.

The cards can be used to test students for any one of the three forms of ESP. *Telepathy*, also called *mental telepathy*, is commonly known as *mind reading*. To test for telepathy, the cards are shuffled and the instructor focuses on the symbol for several seconds while students record the image they think the instructor is viewing. The students record their perceptions (either a drawing or the name of the image) on a sheet of paper, and the instructor reveals the correct order of the symbols after the entire deck is completed.

Clairvoyance is the ability to perceive persons and events that are distant in time or space through the use of ESP. To test for clairvoyance, the cards are shuffled, and then the deck is placed face down. The students record their perceptions for all 25 cards in order, after which the instructor reveals the correct order of the symbols.

Precognition is the ability to perceive an event that has not yet occurred. To test for precognition, the students record their perceptions for all 25 cards in order. The instructor shuffles the cards and then reveals the correct order of the symbols.

Probability for an Individual Participant

The probability of a particular outcome for an individual participant can be determined by computing a *z* score and then using the proportion in the tail of the standard normal distribution. The formula for calculating the *z* score is as follows:

$$z = \frac{X - p(N)}{\sqrt{Npq}}.$$

In this formula, *X* is the number of hits, p is the probability of a hit on any given trial, q is the probability of a miss on any given trial, and *N* is the number of trials. For example, in computing a *z* score for getting 10 or more symbols correct, *X* = 10, p = .2 (the probability of a hit on any trial is one out of five, or 20%), q = .8, and *N* = 25 (because there are 25 cards in the deck). Thus, the *z* score for 10 or more hits is equal to 2.50, and the probability from the standard normal distribution associated with 2.50 is .0062, which can be found in a table included in most introductory statistics textbooks. Because this probability is less than .01, we can conclude that getting 10 or more symbols correct is not likely to occur by chance. In other words, the null hypothesis—that the person does not have ESP—is unlikely to be true, suggesting that the person does have ESP ability.

Although the purpose of the activity is simply to demonstrate how probability is used to make a decision about the null hypothesis, it is important to note that the exact value of the probability changes depending on the number of people being tested. Because the probability that one person will get 10 or more symbols correct is equal to .0062, the probability that one person will not get 10 or more correct is equal to 1 − .0062, or .9938. In addition, the probability that N people will not get 10 or more correct is equal to .9938 to the Nth power. For example, in a class of 20 students the probability that no one will get 10 or more symbols correct is equal to .9938 to the 20th power, or .883. This means that there is a .117 probability (1 − .883) that at least one student in a class of 20 will get 10 or more symbols correct. Using a significance level of .01 (or even .05), we would have to conclude that such an outcome could have occurred by chance. In a class of 30 students the probability that at least one person will get 10 or more symbols correct is equal to .1702; for a class of 40, it is .2202; for a class of 50, it is .2672. Thus, if you use this demonstration in a fairly large class, the odds are good that one or two students will get 10 or more of the symbols correct, leading some students to conclude erroneously that the null hypothesis should be rejected and that they have ESP ability.

ASSESSMENT

When I first began using the activity, I would administer a brief (five- to six-item) satisfaction survey following the completion of the activity. The survey included such statements as "I found the demonstration interesting"; "The demonstration helped me understand how probability is used in hypothesis testing"; and "After participating in this activity, I have a better understanding of statistical significance." The results of this indirect measure of learning suggested that the activity was effective in enhancing students' understanding of these important concepts.

Instructors using the activity for the first time might want to use more direct measures of student learning; for example, a brief quiz designed to assess students' understanding of the basic concepts could be given on the class day before the activity and again on the class day following the activity. Open-ended questions, such as "How does the notion of indirect confirmation apply to hypothesis testing?" and "What does it mean when the results of a study are statistically significant?" can be particularly useful in gauging students' understanding of these concepts.

DISCUSSION

I have used this activity in introductory psychology and statistics classes for many years to demonstrate the role of probability in making decisions about statistical significance in hypothesis testing. In the introductory psychology classes I generally supplement the activity with a brief video clip of the opening scene from the 1984 film *Ghostbusters*, in which Dr. Peter Venkman, played by Bill Murray, uses Zener cards to test two students for telepathy (see http://www.youtube.com/watch?v=fn7-JZq0Yxs). The video also can serve as a starting point for a discussion of ethical issues in behavioral research, as Venkman delivers electric shock to a male student following the student's incorrect (and sometimes correct) responses to the stimuli. In addition, Venkman uses the term *negative reinforcement* incorrectly, which can lead to a discussion of the difference between negative reinforcement and punishment, a distinction that is challenging for many students. Regardless of whether the video clip is used, students report that they enjoy the ESP demonstration and, more important, it seems to enhance their understanding of how probability is used in significance testing.

REFERENCES

Cohen, J. (1994). The earth is round (*p* <. 05). *American Psychologist, 49*, 997–1003. http://dx.doi.org/10.1037/0003-066X.49.12.997

Nickerson, R. S. (2000). Null hypothesis significance testing: A review of an old and continuing controversy. *Psychological Methods, 5*, 241–301. http://dx.doi.org/10.1037/1082-989X.5.2.241

Rhine, J. B. (1934). *Extra-sensory perception*. Boston, MA: Bruce Humphries. http://dx.doi.org/10.1037/13314-000

Trafimow, D., & Marks, M. (2015). Editorial. *Basic and Applied Social Psychology, 37*, 1–2. http://dx.doi.org/10.1080/01973533.2015.1012991

10 USING A TV GAME SHOW FORMAT TO DEMONSTRATE CONFIDENCE INTERVALS

Alexis Grosofsky

It is important for psychology students to be exposed to the concept of confidence intervals, especially given the shift in psychological research away from null hypothesis statistical testing (e.g., Cumming, 2014; Trafimow & Marks, 2015). The activity presented in this chapter helps students understand the idea of a confidence interval using a game show format. Instructors can assess students' understanding by posing a series of questions listed later in the chapter.

CONCEPT

> I find confidence intervals are among the most difficult and aggravating topics to attempt to teach accurately. You can get "close enough" (ironically) with relatively little effort (just show them how to compute a margin of error), but to get students to appreciate exactly what they actually are/mean is incredibly hard. (Straight, 2014)

The above quote sums up nicely what people may think about teaching confidence intervals (CIs). However, CIs are an important concept in statistics and research methods. The sixth edition of the *Publication Manual of the American Psychological Association* (APA, 2010) puts more emphasis on CIs not only by recommending them but also by giving authors examples of how to report them in their manuscripts (Fidler, 2010). CIs are part of the move away from relying solely on null hypothesis statistical testing. Instructors interested in teaching students APA Style for reporting CIs are referred to pages 34, 117, and 140 (for means) of the *Publication Manual*.

CIs were developed by Jerzy Neyman, who published the idea in 1937 (Chiang, n.d.). In defining CIs, Starmer and Starter (2005) suggested that "if the procedure for computing a 95% confidence interval is used over and over, 95% of the time the interval will contain the true parameter value" (p. 36). CIs have two components: (a) the width (an indication of precision) and (b) the percentage of confidence (an indication of degree of uncertainty).

The wider the difference between the lower limit and the upper limit, the less informative the interval is. For example, being 95% confident that the mean test score is between 5% and 95% is not as useful to us as being 95% confident that the mean test score is between 75% and 82%. The best CI is one that gives the highest degree of confidence over the narrowest range of values.

http://dx.doi.org/10.1037/0000024-011

Activities for Teaching Statistics and Research Methods: A Guide for Psychology Instructors, J. R. Stowell and W. E. Addison (Editors)

No special equipment is needed for this activity; however, it may be helpful to have access to the Internet to play YouTube videos.

The "Range Game" from *The Price Is Right* can be used to help students understand how CIs work. The game involves a kind of price-thermometer with a range of price options. A translucent red bar starts at the lowest values and spans $150—an amount between one fifth and one quarter of the price range. The bar slowly ascends, and the contestant must press a button to stop it when the price of a "valuable prize" is within the area covered by the bar. If the contestant successfully stops the translucent bar when it covers the price of the prize, he or she wins the prize. A YouTube video clip of the game (see, e.g., https://www.youtube.com/watch?v=6ORYo6jiuX8) will ensure that all students are familiar with it before moving on.

In the Range Game, the translucent red bar represents a CI, and the price of the prize represents the population parameter. This is, of course, an imperfect representation of CIs in that no sample is used to generate a CI. Also, the width of the "interval" never changes—it is always $150. Whether this would correspond to a 95, 99, or some other percentage CI is unknown. Similar to actual CIs, however, each Range Game simulated CI either does or does not contain the prize/population value, making it similar to null hypothesis significance testing, in which a result either is or is not statistically significant. Despite the fact that such phrases as "a borderline significant trend," "marginally significant," and "slightly significant" are often seen in the literature, statistical significance is in fact a binary decision (Hankins, 2013). The same reasoning applies to CIs; the true population parameter either is, or is not, in the interval.

To help students understand CIs, you will have to explain how variability affects the width of the CI. This is a good time to review the concepts of variability and standard deviation. Once you have reviewed these concepts, you can ask students what width interval (an indication of amount of variability) is most desirable (see the questions listed in the Assessment section of this chapter).

You should show students a video clip of the Range Game to ensure that they are all familiar with it. Each video clip takes fewer than 3 minutes to show. Once students have a concrete example to help them understand the concept of CIs, you can discuss the two components (width and percentage). Ask students to point out how the Range Game falls short as an example of CIs (see the list of URLs below). Pose discussion questions to the class (to get students to consider factors that will influence the size of the CI). The question-and-answer session should take 15 to 20 minutes.

Below are links to four video clips that illustrate various outcomes of the Range Game. Duration of the clips ranges from 1:56 to 2:25:

- https://www.youtube.com/watch?v=AOVlo3vduao—prize price is higher: loser
- https://www.youtube.com/watch?v=IooA0guouuI—prize price in "range": winner
- https://www.youtube.com/watch?v=cqDxt6B5Zkw—prize price in "range": winner
- https://www.youtube.com/watch?v=elHU6IfqqWw—prize price is lower: loser

Listed below are some of the questions you can pose to students, along with possible answers shown in italics (Grosofsky, 2014):

- Where does the Range Game fall short in portraying CIs? *No sample is used to calculate it. We do not know the percentage of confidence. Every range is identical.*

- From the contestant's perspective, what width of the translucent red bar would be best? Why? *A bar that spanned all of the possible prices would be best because then the contestant would win every time.*
- From the contestant's perspective, what information would increase the odds of winning with the game as configured? *Guessing the price of an item you are familiar with would help as you would have a better idea of the typical price and the variability around it (range of low and high prices for other versions of the item).*
- From the perspective of the show's producer, what width of red bar would be best? Why? *A red bar that is very narrow would be the best because very few prizes would be awarded, thus saving the show money. (Or) A red bar that is the same size as it is now would be the best because a reasonable number of prizes would be awarded, encouraging people to want to be contestants and, hopefully, encouraging viewers to tune in.*
- What would happen to the rate of prizewinning if the bar was much narrower? Why? *Fewer prizes would be won. The narrower the interval, the harder it would be for the price of the prize to fall in the interval.*
- What would happen to the rate of prizewinning if the red bar was much wider? Why? *More prizes would be won. The broader the interval, the easier it would be for the price of the prize to fall in the interval.*
- Moving away from the Range Game to CIs in general: Because we can be more confident about containing the parameter with larger intervals, shouldn't all CIs be as large as possible? *No, a very broad range will not be meaningful (refer to mean test score example above). Note that there is a tradeoff between degree of certainty and width of the CI. In general, the more certain you want to be, the wider (larger) the interval needs to be. The goal is to have a high level of confidence paired with a small interval.*
- More CIs in general: How can this information be used in the "real world?" For these examples, consider what having more (or less) variability in your sample would mean. *You can calculate CIs for any numerical value of interest; for example, amount of time spent daily on social media sites, amount of money spent at restaurants monthly, number of each size of a clothing item a store should have in stock, understanding political polls, and determining cutoff scores for psychometric instruments (e.g., to label severity of autism spectrum disorder). In all cases, larger variability (perhaps due to smaller sample sizes) would result in wider (therefore less informative) CIs. The opposite would be true for smaller variability.*
- How could variability lead to a smaller interval? *If you have little variability in your sample, you will have a narrower interval. Another way is to have a narrower interval is to have little variability in your population. (The instructor may wish to remind students that larger samples are more representative of populations.)*

Questions similar to these can be developed to use as test items.

DISCUSSION   This activity should help students understand the basics of CIs. It is important to emphasize that the CI is generated from a sample and is about a population parameter. Sometimes students mistakenly believe that the CI is about sample statistics rather than parameters.

ADDITIONAL RESOURCES

Baggett, J., Fridley, B., & Reineke, D. (2011, August 4). *Interpretation of confidence intervals*. Retrieved from https://lessonstudyproject.wordpress.com/2011/08/04/mathematics-interpretation-of-confidence-inte-43582/

Cumming, G., Williams, J., & Fidler, F. (2004). Replication and researchers' understanding of confidence intervals and standard error bars. *Understanding Statistics*, 3, 299–311. http://dx.doi.org/10.1207/s15328031us0304_5

Dempsey, M. (2013). Dealing cards with confidence. *The Mathematics Teacher*, 106, 459–463. http://dx.doi.org/10.5951/mathteacher.106.6.0459

Gould, R., Rumsey, D., Miller, G., Patel, R., Snell, M., & Sullivan, M. (n.d.). *Schematyc: Unit 10. Confidence intervals*. Retrieved from http://schematyc.stat.ucla.edu/unit_10/teaching_tips.php

Kalinowski, P. (2010). Student Notebook: Understanding confidence intervals (CIs) and effect size estimation. *Psychological Science*, 23(4). Retrieved from http://www.psychologicalscience.org/index.php/publications/observer/2010/april-10/understanding-confidence-intervals-cis-and-effect-size-estimation.html

Rice Virtual Lab in Statistics. (2008). *Confidence interval simulator*. Retrieved from http://www.ruf.rice.edu/~lane/stat_sim/conf_interval/

REFERENCES

American Psychological Association. (2010). *Publication manual of the American Psychological Association* (6th ed.). Washington, DC: Author.

Chiang, C. L. (n.d.). *Statisticians in history: Jerzy Neyman 1894–1981*. Retrieved from http://www.amstat.org/about/statisticiansinhistory/index.cfm?fuseaction=biosinfo&BioID=11

Cumming, G. (2014). The new statistics: Why and how. *Psychological Science*, 25, 7–29. http://dx.doi.org/10.1177/0956797613504966

Fidler, F. (2010). The *American Psychological Association Publication Manual, Sixth Edition:* Implications for statistics education. In C. Reading (Ed.), *Data and context in statistics education: Towards an evidence-based society. Proceedings of the Eighth International Conference on Teaching Statistics (ICOTS8, July, 2010), Ljubljana, Slovenia*. Voorburg, the Netherlands: International Statistical Institute. Retrieved from http://iase-web.org/documents/papers/icots8/ICOTS8_C156_FIDLER.pdf

Grosofsky, A. (2014, January 3). *Illustrating confidence intervals with a game show*. Retrieved from http://www.teachpsychscience.org/resource.asp?id=137

Hankins, M. (2013, April 21). Still not significant: Probable error: I don't mean to sound critical, but I am; so that's how it comes across [Web log post]. Retrieved from https://mchankins.wordpress.com/2013/04/21/still-not-significant-2/

Starmer, J., & Starter, F. (2005, May 19). *The joy of learning. Main ideas, scaffolding, and thinking: Building new concepts by modeling: HOWTO*. Retrieved from https://frank.itlab.us/datamodel/

Straight, N. (2014, January 23). Re: Games for teaching probability #6: Confidence intervals [Online forum discussion message]. Retrieved from https://boardgamegeek.com/thread/1109296/games-teaching-probability-6-confidence-intervals (Comment at: https://boardgamegeek.com/article/14640529#14640529)

Trafimow, D., & Marks, M. (2015). Editorial. *Basic and Applied Social Psychology*, 37, 1–2. http://dx.doi.org/10.1080/01973533.2015.1012991

11 REAL-LIFE APPLICATION OF TYPE I AND TYPE II DECISION ERRORS
Bernard C. Beins

The activity described in this chapter introduces students to scenarios that help them distinguish between Type I and Type II decision errors in research and understand why the context determines which error is more serious.

CONCEPT

In the long run, research supported by statistical arguments is likely to lead to valid conclusions, but the possibility of erroneous decisions in any single study always exists. Researchers know that they have to live with a chance of making a mistake in their decisions because of random, unpredictable fluctuations in the data or to methodological flaws in the research.

Researchers begin by formulating a hypothesis that there is nothing interesting going on with respect to the factors they are studying (e.g., no difference between group means, no relationship between variables). This is the *null hypothesis* (H_0) that they hope to reject. Their real belief, expressed as the alternative, or *research hypothesis* (H_1), is that something interesting really is happening (e.g., there is a difference between group means, there is a relationship between variables).

The norm among researchers in psychology is to accept a difference between groups or a correlation as statistically significant if the result, or a more extreme result, would occur less than 5% of the time when there really is no effect. This is the alpha level, which is also the Type I error rate. It is also possible that researchers may conduct a study and erroneously conclude that there are no significant differences between groups or no significant relationships among variables. This is a Type II decision error; researchers typically do not know how often they commit this kind of error.

Table 11.1 illustrates the relation between the decision an investigator makes and the true state of affairs, which we never really know because we always have incomplete information (i.e., we can only make inferences based on the data from our sample).

The activity described in this chapter uses an example that should be of interest to many psychology majors even though it involves a medical topic: breast cancer detection. To start, you should refresh students' awareness that researchers begin by assuming that the null hypothesis is true and that all decisions are made with respect to the null hypothesis (i.e., reject or do not reject the null hypothesis). It is important to make sure that students know that there is always the possibility of error because tests are not always right. Sometimes the tests miss the cancer, or sometimes they falsely indicate the presence of cancer.

http://dx.doi.org/10.1037/0000024-012

Activities for Teaching Statistics and Research Methods: A Guide for Psychology Instructors, J. R. Stowell and W. E. Addison (Editors)

Table 11.1   *Table of Decisions and Their Relation to the True Situation*

	The decision	
The truth	Nothing is going on (Do not reject H_0)	Something is going on (Reject H_0)
Nothing is going on	Research: Correct inference Medicine: Negative outcome	Research: Type I error Medicine: False positive
Something is going on	Research: Type II error Medicine: False negative	Research: Correct inference Medicine: Positive outcome

Note. H_0 = null hypothesis. Copyright 2015 by Bernard C. Beins. Adapted with permission.

MATERIALS NEEDED

The only materials needed for this activity are the handouts provided in Appendixes 11.1 and 11.2. Most of the actual activity involves student discussion of which type of possible error has more serious negative consequences.

INSTRUCTIONS

After providing students with a copy of the decision matrix in Table 11.1, you should present the following questions (answers are in italics):

1. What is the null hypothesis in this example? *There is no cancer.*
2. What does the doctor conclude if the null hypothesis is true? *There is no cancer.*
3. What is the decision if the doctor rejects the null hypothesis? *There is cancer.*
4. What constitutes a Type I decision error? *Concluding that the woman has breast cancer when she does not.*
5. What constitutes a Type II decision error? *Concluding that the woman does not have breast cancer when she does.*

After providing students with a copy of Appendix 11.1 (without the answers), you should emphasize that the actual base rate of cancer in this age group is 1% and the rates of Type I and Type II error are 18% and 10%, respectively. Then ask students to compute the number of women who fall into each category of correct and incorrect decisions.

As Appendix 11.1 reveals, in a sample of 1,000 women 40 years of age, many who do not have cancer will test positive for it. In such a sample, 178 women will be told erroneously that they have cancer; of the 10 women who actually have cancer, nine women will test positive. In other words, a total of 187 women will test positive. Thus, what percentage of women who test positive actually have cancer? The answer is nine out of 187, or approximately 4.8%. The vast majority of women in this age group who test positive for breast cancer do not have it.

How important are Type I and Type II decision errors? They can be critical in some circumstances. In the case of breast cancer, which type of error is worse (i.e., has more serious negative consequences)? This seems like an easy question to answer. It would seem at first glance that a false negative is worse because the cancer is missed, so the health outcomes of patients in this category are seriously at risk.

However, the answer is not altogether obvious. Women who experience false positive results show some short-term psychological distress and have to undergo additional screening tests, which can be costly. These additional tests are also not perfect and could lead to potentially dangerous treatments. False positives cost the economy about $4 billion a year in the United States, so some medical groups have recommended against yearly mammograms for younger women because cancers are so rare (Ong & Mandl, 2015).

The decision about mammograms requires a judgment that the statistics cannot resolve. However, knowledge of the likelihood of Type I and Type II errors and their ramifications will be useful in helping people make their decision.

DISCUSSION

Students become engaged in this exercise because breast cancer has affected many of them, either directly or indirectly. Thus, the activity has a practical focus that is meaningful to them. It would be prudent to inform students ahead of time about the nature of the topic because their own personal experience with breast cancer might make them uncomfortable.

ASSESSMENT

Appendix 11.2 presents a scenario that is useful for testing student knowledge. It asks them to identify the two error types, what decision would be associated with a specified score, and which type of error might have more negative consequences for a person who has undergone this test.

ADDITIONAL RESOURCE

McGrayne, S. B. (2011). *The theory that would not die: How Bayes' rule cracked the enigma code, hunted down Russian submarines, and emerged triumphant from two centuries of controversy.* New Haven, CT: Yale University Press.

REFERENCE

Ong, M. S., & Mandl, K. D. (2015). National expenditure for false-positive mammograms and breast cancer overdiagnoses estimated at $4 billion a year. *Health Affairs, 34,* 576–583. http://dx.doi.org/10.1377/hlthaff.2014.1087

BERNARD C. BEINS

Appendix 11.1

Probabilities and Numbers of Different Types of Correct and Erroneous Decisions in Detecting Breast Cancer for 40-Year-Old Women

The table is calculated from a sample of 1,000 women. The base rate of breast cancer in this age group is 1%. (Answers are shown in italics.)

	The decision		
The truth	No cancer (Do not reject H_0)	Cancer (Reject H_0)	
No cancer	Correct negative 82% $0.82 \times 990 = 812$	False positive Type I error 18% $0.18 \times 990 = 178$	Actual cases of no cancer $812 + 178 = 990$
Cancer	False negative Type II error 10% $0.10 \times 10 = 1$	Correct positive 90% $0.90 \times 10 = 9$	Actual cases of cancer $9 + 1 = 10$
	Diagnosed with no cancer $812 + 1 = 813$	Diagnosed with cancer 187	Total sample 1,000

Note. H_0 = null hypothesis. Copyright 2015 by Bernard C. Beins. Adapted with permission.

Appendix 11.2

Assessment Exercise for Students

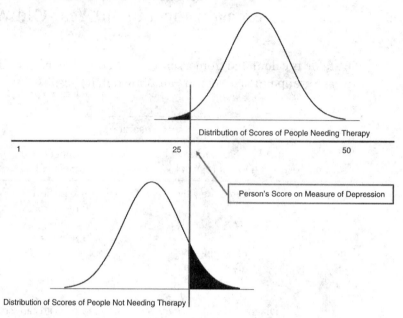

Note. Copyright 2015 by Bernard C. Beins. Reprinted with permission.

ASSESSMENT QUESTIONS

1. What conclusions would represent a Type I decision error and a Type II decision error regarding whether or not a person needs therapy?
   - *Type I decision error: Concluding that a person needs therapy but does not.*
   - *Type II decision error: Concluding that the person does not need therapy, but the person actually does.*
2. Is it more likely that a person with a score of 27 would come from the population of people who need therapy or from those who do not need it? *There are more people from the population not needing therapy who score above 27 than there are people from the population needing therapy who score lower than that value. There will be more Type I than Type II errors.*
3. Which type of decision error would be more serious? *If a person experiences depression, it would be problematic if the individual did not receive treatment. On the other hand, if somebody did not need the therapy, being diagnosed as depressed could lead to social stigma, could cost the person money for therapy, and would waste the time of both the therapist and the client. This is a judgment call.*

12 Factors That Influence Statistical Power

Michael J. Tagler and Christopher L. Thomas

The purpose of the activity introduced in this chapter is to help students understand the concept of statistical power by having them collect and analyze actual data; interpret their results (effect size and statistical significance); and explore how sample size, effect size, and significance criterion influence power.

CONCEPT   In null hypothesis significance testing (NHST), *power* is the probability of rejecting the null hypothesis when it is false (Cohen, 1992). In other words, power is the conditional probability of correctly detecting a statistically significant result (a *true positive*). The probability is conditional because it requires the assumption that the null hypothesis is false—that an effect of a particular magnitude (effect size) truly exists in the population. Power is symbolized as $1 - \beta$, given that it is the complementary outcome to a Type II decision error (i.e., β; the probability of retaining a false null hypothesis, also known as a *false negative*).

Sample size, error variability, effect size, and significance criterion are the major influences on power. Because increasing sample size increases power, statistically significant results are more likely with larger samples. Greater error variability decreases power, making it more difficult to obtain significant results. Power is greater when the effect under investigation is strong. For example, a large difference between a sample and population mean is more likely to be statistically significant than a small difference. Moreover, selecting a less stringent significance level (e.g., $\alpha = .05$ instead of $\alpha = .01$) makes it easier to reject the null hypothesis. Although the significance criterion influences power, the value of α describes a different conditional probability: the probability of rejecting a true null hypothesis (also known as a *false positive*, or a *Type I decision error*). Finally, $1 - \alpha$ is the probability of correctly failing to reject the null hypothesis (a *true negative*).

The concept of power as a probability is somewhat distinct from the procedure of power analysis. Ideally, researchers should conduct power analyses before collecting data to determine the minimum sample size needed to conduct a study with sufficient power. To do so, the researcher must specify the minimum population effect size theorized to be important to their investigation, set the α level, and select the desired level of power.

In contrast to a priori power analyses, researchers often inappropriately conduct power analyses after collecting and analyzing their data using the sample (observed) effect size. For example, a researcher who is disappointed to find that his or her study did not yield statistically significant results may then run a power analysis with the goal of determining whether the study was powerful enough to detect a significant

http://dx.doi.org/10.1037/0000024-013
Activities for Teaching Statistics and Research Methods: A Guide for Psychology Instructors, J. R. Stowell and W. E. Addison (Editors)

effect. However, because post hoc power is computed on the basis of the sample effect size, there is always a 1:1 inverse relationship between observed power and p values. To be specific, higher (not statistically significant) p values always correspond to lower observed power (Hoenig & Heisey, 2001).

Despite the caution against conducting post hoc power analyses in actual research contexts, we suggest it is useful for statistics students to examine the observed power of an already-conducted statistical test because it provides a full story for the student to decipher. For example, if low power is observed, the student can then logically examine and make judgments regarding the degree to which sample size, effect size, and/or α level were contributing factors. With this perspective in mind, the class activity we describe here takes a retrospective approach to power because it is designed to demonstrate the influences of sample size, effect size, and α level.

MATERIALS NEEDED

Technology

The instructor should provide students with a simple method to compute power in a one-sample z-test scenario in which an obtained sample mean is compared to a known population mean. The following values are needed to carry out a one-sample z test: population mean, population standard deviation, sample mean, sample size, and α level.

Computer software is needed to make the computations quickly. Depending on existing statistical knowledge and technical skill, a great many options exist, including G*Power (http://www.gpower.hhu.de/en.html) and formulas in Microsoft Excel (Black, 1999). However, we recommend the R script available here: http://www.r-fiddle.org/#/fiddle?id=3YrAbJI8. The R Project (https://www.r-project.org) is open-source software that provides a free alternative to traditional statistical packages such as SPSS and SAS. However, installation of and learning to use R likely requires more time than allowed for a class activity. We recommend instead using a web-based R console such as R-Fiddle (http://www.r-fiddle.org). R-Fiddle is ideal for the classroom because it does not require installation; it runs in a web browser, including those on mobile devices; it requires very little instruction to get started; and it does not require user registration. Because students will work in pairs or small groups for this activity, it is also not essential that each student have his or her own Internet device.

Self-Report Measure

The instructor distributes to students a brief self-report measure to generate the data for the activity. Any number of instruments would be appropriate, but we recommend selecting a measure with established norms that is simple to complete and score and that does not assess sensitive information (e.g., sexual behavior, drug use, psychopathology).

We use the Life Orientation Test—Revised (LOT–R; Scheier, Carver, & Bridges, 1994). The LOT–R is brief, easy to score, and is a widely used measure of optimism. The LOT–R and pertinent scoring information is available in Table 6 of Scheier et al. (1994). On the basis of the results from more than 2,000 college students, Scheier et al. (1994) reported the following normative information for the LOT–R: $M = 14.33$, $SD = 4.28$. These values can be used as the population parameters for the one-sample z test, as we describe below.

Students begin the activity individually by completing and scoring the LOT–R. Depending on the class size, you should divide students into groups or pairs and have them calculate the mean LOT–R score for their group or pair. Next, have students conduct a one-sample z test in which they compare their group mean with the population mean using Scheier et al.'s (1994) norms; they should also determine the power of their test. We suggest the instructor distribute the R-script in a text file to students, or simply use the "Save" option in R-Fiddle to create a custom hyperlink with the R script already entered into the R editor (which can be seen at the top of screen in the R-Fiddle script). We have saved the script at the following site: http://www.r-fiddle.org/#/fiddle?id=3YrAbJI8. Before running the script, students need to type in values for the population mean, population standard deviation, sample mean, sample size, and α in the "Characteristics of Data" section (lines 2–6). The population mean and standard deviation from Scheier et al. are already entered in the R script, along with an example of a sample mean ($M = 17.00$) and a sample size ($n = 3$).

When the R script is run, the console (bottom of screen) displays the test results, effect size (Cohen's d), sample size, and observed power. You can initiate a class discussion focusing on the findings of each group by prompting students to share their results. For instance, using the example values ($M = 17.00$, $n = 3$, $\alpha = .05$) preentered in the R script yields a medium effect size ($d = 0.62$) but an underpowered (0.19), statistically nonsignificant result ($p = .28$). A group that obtains these results should logically conclude that the small sample size is the culprit for the nonsignificant, low-power result.

Next, you should have students repeat the activity with larger sample sizes by instructing several of the groups/pairs to merge. We suggest repeating the activity until students have merged to create a single group composed of the entire class. Unless combining groups results in dramatic effect size changes, power should increase with the larger groups. For the final classroom discussion we recommend describing hypothetical situations to solidify students' understanding of the factors that affect power; for example, you may ask students how power would be affected by even larger sample sizes or a more stringent α level and then demonstrate the results using the R script.

ASSESSMENT You can tailor assessment of learning to fit the particular course. Worksheets can easily be created in which research scenarios are presented and students use the R script to document the factors that influence power. We suggest exposing students to a variety of scenarios to deepen their understanding (e.g., small and large effect sizes, small and large sample sizes, statistically significant and nonsignificant outcomes). A quick but informative assessment is to have students take the last 5 minutes of class to write what they learned from the activity, what they may not fully understand, any remaining questions they have, and whether they thought the activity was helpful.

DISCUSSION A major challenge that arises in this activity is that there are multiple, interrelated concepts that students must consider simultaneously to fully understand the concept of power. These include probability, null and alternative hypotheses, steps involved in NHST, four possible outcomes in NHST, how p values are correctly interpreted, and how measures of effect size provide information that is not confounded with sample size. All of this can be overwhelming to new students of statistics, so we urge you to be patient and provide students enough time and practice to work with these concepts.

There are certainly additional details that can be covered in classroom activities devoted to power. For example, we focus on conducting two-tailed (nondirectional) tests, but an instructor could include an examination of one-tailed versus two-tailed tests (one-tailed tests are more powerful). We also chose to simplify the activity by using a one-sample z test, but one could revise the scenario by using a one-sample t test, so that the sample standard deviation influences the results. Moreover, instructors can initiate a broader discussion of how research design issues can influence statistical power. For example, you can ask students to consider how the choice of a stronger versus weaker manipulation (e.g., a large drug dose vs. a small drug dose), or the reliability of dependent measures, is likely to influence power.

REFERENCES

Black, T. R. (1999). Simulations on spreadsheets for complex concepts: Teaching statistical power as an example. *International Journal of Mathematical Education in Science and Technology, 30,* 473–481. http://dx.doi.org/10.1080/002073999287752

Cohen, J. (1992). Statistical power analysis. *Current Directions in Psychological Science, 1,* 98–101. http://dx.doi.org/10.1111/1467-8721.ep10768783

Hoenig, J. M., & Heisey, D. M. (2001). The abuse of power: The pervasive fallacy of power calculations for data analysis. *The American Statistician, 55,* 19–24. http://dx.doi.org/10.1198/000313001300339897

Scheier, M. F., Carver, C. S., & Bridges, M. W. (1994). Distinguishing optimism from neuroticism (and trait anxiety, self-mastery, and self-esteem): A reevaluation of the Life Orientation Test. *Journal of Personality and Social Psychology, 67,* 1063–1078. http://dx.doi.org/10.1037/0022-3514.67.6.1063

13 AN INTERDISCIPLINARY ACTIVITY FOR p VALUES, EFFECT SIZES, AND THE LAW OF SMALL NUMBERS

Andrew N. Christopher

■ ——————————————————————————————— ■

Using archival performance data of stocks and bonds, this activity can be used to illustrate probability values and effect sizes, the difference between these two concepts, and the law of small numbers.

■ ——————————————————————————————— ■

CONCEPT

This activity not only allows students to learn about p values, effect sizes, and the law of small numbers but also, in the process, it facilitates the development of basic knowledge and skills that any educated person should possess. In addition to learning about the difference between p values and effect sizes in reporting inferential statistics, students who complete this activity learn about the relationship between investment risk and return as well as the concept of compounding returns and how they are calculated. Furthermore, the activity can serve as a discussion point about the concept of variability (or, in investment language, *volatility*). Finally, the activity can be used to facilitate a discussion of the *law of small numbers*, which is the notion that atypical results are more likely to occur with a small sample of observations relative to a large sample.

MATERIALS NEEDED

You will need a list of annual returns for one stock index and one bond index for a prolonged period of time. I generally use the annual returns of the Standard & Poor's 500 (S&P 500; stocks) and the 10-year Treasury Bond dating back each year to 1928 (see Damodaran, 2015). Because I teach relatively small classes (about 24 students), I write each pair of annual returns on precut index cards. These cards are placed into a small box or other container (e.g., hat), from which they will be drawn out during the activity.

INSTRUCTIONS

I use this activity immediately after teaching students how to hand-calculate an independent-samples t statistic, as a lead-in to understanding the p value and effect size concepts that accompany the t test. Thus, students have presumably learned the basics of hypothesis testing, including the notion of p values, prior to participating in this activity.

In class, the activity is conducted in two parts. In the first part, each student is given \$10,000 of hypothetical money to invest. Students have only two investment choices: (a) stocks (as measured by the S&P 500) and (b) bonds (as measured by the

Laura Brandt offered valuable feedback on previous drafts of this chapter.

http://dx.doi.org/10.1037/0000024-014

Activities for Teaching Statistics and Research Methods: A Guide for Psychology Instructors, J. R. Stowell and W. E. Addison (Editors)

10-year Treasury). I tell students they will need the money in 3 years to make a down payment on a house. They can divide their $10,000 however they wish between the two options, and I inform them that history shows that, over the long term, stocks tend to be better investments than bonds. We then draw one card out of the box (or container) that contains an annual return for both the S&P 500 and the 10-year Treasury. On the basis of how much money they allocated for each investment, students need to calculate how much money they now have in that investment after the first year. It is advisable to show them how to perform this calculation. For example, if they have $5,000 in stocks, and this investment increases in value by 8%, they need to multiply $5,000 by 1.08 to determine how much money they now have ($5,400). Such calculations become increasingly important to understand as we move into subsequent years in the demonstration and the effects of compound investment returns kick in.

Once students have determined how much they have in stocks and bonds after Year 1, I pull another card out of the box. On the basis of these Year 2 returns, students calculate how much money they have in their account. You will need to make sure they are using the money they calculated at the end of Year 1 as the starting point for calculating how much their investments are worth after Year 2. Once students have calculated how much money they have in stocks and bonds, I draw a card with our Year 3 returns, and students calculate the resulting value of their investments.

Before conducting the second part of the activity, you should point out that 3 years is a relatively short time period when it comes to investments. Students restart the activity with $10,000 to invest in stocks and bonds. However, rather than needing the money in 3 years, this time the money will not will not be needed for 25 years, when they will make a down payment on a dream vacation home. They can again divide their $10,000 between stocks and bonds as they desire. Once students initially allocate their money between stocks and bonds, we draw a card with the return for each investment 25 times, each time constituting 1 year of returns. After seeing each year's return, students calculate how much money they have in each investment.[1]

After completing both the 3- and 25-year investment time horizons, students calculate an independent-samples t test[2] and compare the average annual returns, p values, and effect sizes for each investment time horizon. With the 3-year horizon it is relatively easy to perform these calculations by hand, or put the data into an Excel or SPSS spreadsheet, perform the test, and interpret the output these programs provide.

In addition, my students and I discuss the law of small numbers to explain why the 3-year horizon is more likely to provide "weird" results, which in this context would occur when bonds outperform stocks. It could be that if we drew a year in which bonds performed better than stocks by a wide margin (e.g., 2008), our overall results for the relatively brief time horizon would be unusual by historical standards. However, with a 25-year horizon such aberrations are more likely to be offset by historically "normal" investment returns.

---

[1] To make the demonstration more realistic, you can allow students to rebalance their investment choices after each year of the demonstration.

[2] Because the correlation between the returns of the S&P 500 and 10-Year Treasury is trivial ($r = -.03$), they are statistically independent of each other.

Pre–post analyses for students who have completed this activity suggest that it could be effective in helping them understand situations in which conclusions are based on small sample sizes and the notion of p values as used in research. For instance, consider the following exam question:

> I have been challenged to a basketball shooting contest by LeBron James, who is arguably the best basketball player in the world. He is allowing me to set the rules for our contest. Based on what you've learned in this class, should I make the contest a best out of 1 shot, or a best out of 20 shots? Explain your answer based on information you learned in this class.

At the beginning of the class, approximately 62% of the students answered the question correctly, but only about 25% of these students could explain the logic of a 1-shot contest. By the end of the semester, every student could identify the scenario with a better chance to defeat LeBron James. More important is that more than 80% of the students could explain it within the context of the law of small numbers.

Furthermore, it appears that students improve their ability to understand the notion of compounding returns. For instance, at the beginning and end of the semester, students were asked the following open-ended question:

> Suppose you invest $100 for 2 years. During the first year, your investment loses 10% of its value. During the second year, your investment gains 10% of its value. How much money do you have at the end of the second year?

At the beginning of the semester, almost 80% of the students answered the question incorrectly, with almost all of these incorrect responses being "$100." At the end of the semester, more than 80% of the students correctly answered the question (i.e., $99).

Other samples questions that could be asked as part of a pre–post assessment include the following two questions:

1. A statistically significant outcome is defined as an outcome that has a _____ probability of occurring if the _____ hypothesis true.
   a. small; null
   b. small; research
   c. large; null
   d. large; research
2. What information does an effect size provide over and above the information provided by a p value in a hypothesis test?

For the 3-year investment window, a relatively small investment time frame, you should expect almost anything in terms of the comparison between stock and bonds returns. Twice when I have performed this activity with my students bonds actually outperformed stocks, although, because of the small sample size, there was not a statistically significant difference. However, according to Cohen's (1992) guidelines it is possible to obtain a moderate effect size of type of investment on returns obtained using a 3-year investment horizon.

For the 25-year demonstration, overall mean stock returns have been better than overall bond returns, something I verify via an independent-samples t test in class using computer software. Of course, stock returns have also been more variable than bond returns, again allowing instructors to point out small periods of time in which aberrant investment returns may be obtained. It is possible that if stock returns toward the end of the 25-year period are particularly bad, this could result in bonds outperforming stocks for the entire time period. If this is the case, however, it would provide another opportunity to highlight the law of small numbers. In addition, such an occurrence would not hamper the calculation and discussion of effect size for type of investment or the notion of p values.

REFERENCES

Cohen, J. (1992). A power primer. *Psychological Bulletin, 112,* 155–159. http://dx.doi. org/10.1037/0033-2909.112.1.155

Damodaran, A. (2015). *Annual returns on stock, T. Bonds and T. Bills: 1928–current.* Retrieved from http://pages.stern.nyu.edu/~adamodar/New_Home_Page/datafile/ histretSP.html

II
RESEARCH METHODS

14 AN ACTIVITY FOR TEACHING THE SCIENTIFIC METHOD

R. Eric Landrum

Science is a process for studying and understanding the world, but it also yields information with unique characteristics; that is, science is both a method and a product. The goal of the activity presented in this chapter is for students to enhance their understanding of the scientific method by first reviewing the steps of the scientific method and their relation to psychological research and then applying concepts of the scientific method and product of science to popular psychological myths. This activity can be used with students individually or in a group setting to encourage collaboration and team-building skills. A pre- and postactivity assessment procedure is recommended so that students and instructors can observe how thinking about the scientific method can change through deep engagement with the concepts.

CONCEPT

What is the scientific method? The word *science* comes from the Latin *scientia*, meaning "knowledge." A formal definition from the American Psychological Association (APA; 2014) is that the scientific method is "a set of procedures, guidelines, assumptions, and attitudes required for the organized and systematic collection, interpretation, and verification of data and the discovery of reproducible evidence, enabling laws and principles to be stated or modified" (p. 327).

However, the word *science* and the application of the scientific method have multiple implications. First, science is a method of studying the world around us, a concept that is taught early in formal education (see the left side of Figure 14.1 for a brief reminder). The scientific method is used in chemistry, physics, biology, and psychology. The right side of Figure 14.1 displays how psychologists translate the traditional five-step scientific method for use in research articles, which may include laboratory reports, student research papers, and even formal manuscripts submitted for publication to peer-reviewed journals.

Second, science produces scientific knowledge that can be distinguished from nonscientific knowledge by the method used to generate it. For example, TV advertisements, news broadcasts, or Internet blogs may report findings that are based on subjective opinion or informal survey research, producing results that can differ in method and outcome from a scientific study. Knowledge gained through a scientific approach has certain characteristics that are different from knowledge gathered in nonscientific ways; these characteristics include greater objectivity; evidence-based (empirical) observation; reliable measurement; and clear, operational definitions. Science is both a method and a product.

http://dx.doi.org/10.1037/0000024-015

Activities for Teaching Statistics and Research Methods: A Guide for Psychology Instructors, J. R. Stowell and W. E. Addison (Editors)

Steps of the scientific method	Writing in psychology
Generate a problem →	Introduction
Form a hypothesis →	
Conduct an experiment →	Method
Collect data →	Results
Draw a conclusion →	Discussion

Figure 14.1.   The scientific method and psychology.

MATERIALS
NEEDED

Please see the example handout in Appendix 14.1. Although a printed paper version will work, the activity is conducted more easily electronically (e.g., using a Google Form, Google Doc, or some other collaborative online tool). If a paper form is used, you should modify the handout to include extra space after each question for students' responses.

INSTRUCTIONS

Preactivity Assessment

You should have students respond individually to the two essay questions below, without the aid of electronic sources, classmates, or other type of reference material:

1. What are the steps of the scientific method? Describe each of these standard steps and how they relate to psychological research studies.
2. In thinking about the scientific method in psychological research, apply the ideas of "science as a method" and "science as a product" to a common claim heard today. Outline the steps you would follow to explore, scientifically, whether the claim had merit or not.

Activity

In Part 1 of the activity, have students define each step of the scientific method. They should consult with classmates and/or electronic sources to identify the steps and then match those steps with the typical portions of an APA Style–formatted research paper (i.e., Introduction, Method, Results, and Discussion).

In Part 2 of the activity, students should, while working in teams designated by you, complete the table in the handout (see Appendix 14.1), one claim ("row") at a time. As they do so, students should discuss their responses as a team and describe some of the specific methodological choices to be made when testing the claim using scientific research methods. After envisioning what the methodology might look like for the claim, students should describe what the poststudy scientific "product" might look like (i.e., the outcome).

ASSESSMENT

To assess students' mastery of the material, ask them again to individually complete the two preactivity assessment questions and to compare their answers before and after the activity. You also can analyze differences between students' pre- and postactivity assessment responses.

DISCUSSION

This activity should generate lively discussions in small groups. As time permits, discussions can be lengthy (the entire class period), or they can be more focused on one particular aspect of the scientific method. Parts of the activity can be revisited over time and used thematically over the duration of the course. The skills for this activity are within the capabilities of most psychology majors but difficult enough that students will benefit from collaboration. Students may be unclear on their recollections of the steps of the scientific method (depending on whether these steps have been discussed in class recently), but the steps should be clear by the end of the activity. Some students may want to delve deeply into one topic while not addressing others. You should think about timing and specific goals, such as "Should students complete a set number of scenarios, or should they self-pace through the exercise in a given amount of time?" Ultimately, you will need to guide students through the exercise at the level of detail desired.

Postactivity exercises can include researching the actual claims and the current data available to support or refute them and discussing the appropriate conclusion/scientific product. For a follow-up exercise, students can explore the Internet to identify additional psychological myths and continue to add to the list of claims in the current handout; alternatively, they could generate their own topics for further exploration. This exercise would help keep the discussion topics current over time.

REFERENCES   American Psychological Association. (2014). *APA dictionary of statistics and research methods*. Washington, DC: Author.

Standing, L. G., & Huber, H. (2003). Do psychology courses reduce belief in psychological myths? *Social Behavior and Personality, 31*, 585–592. http://dx.doi.org/10.2224/sbp.2003.31.6.585

Appendix 14.1

Assessment Exercise to Help Students Learn About the Scientific Method

For each of the claims listed below, think about the steps of the scientific method that would be used to research that claim (conduct an experiment, collect data, draw conclusions) and what a likely product (outcome) would look like. For example, if the claim were made that "UFOs exist," what scientific steps could be used to explore that claim? (Describe an experiment you could conduct, including how you would collect data and how you would draw conclusions from that data.) Based on a scientific approach and the use of research methods, what might be a likely scientific product (outcome) of that research?

Claims ↓	Science as a method			Science as a product
	Conduct an experiment	Collect data	Draw conclusions	
People use only about 10% of their brain's capacity.	Learn about fMRI, PET, and CAT scans and figure out which one is a good indicator that an area of the brain is "in use."	Ask individuals to be tested under different conditions (e.g., at rest and while reading) using the selected brain scan technology.	Calculate total area of brain activity under different conditions and compare that number to 10%.	Share the results publicly, such as on a website or blog.
The world's most popular drug is alcohol.				
During a full moon, people commit more crimes and behave more abnormally.				
A few people can use their minds to influence, for example, the way that dice will fall.				
Inside our brains are memories for everything we have ever experienced.				
As people age, they sleep more.				
Through hypnosis, some people can remember things from the first 6 months of life.				

Claims ↓	Science as a method			Science as a product
	Conduct an experiment	Collect data	Draw conclusions	
Clear evidence exists to show that a very small percentage of people can receive the thoughts of others and predict the future.				
Mentally ill individuals are more likely to be violent than normal individuals.				

Note. These claims are from Standing and Huber (2003). fMRI = functional magnetic resonance imaging; PET = positron emission tomography; CAT = computerized axial tomography.

15

Linking Identification of Independent and Dependent Variables to the Goals of Science

Mary E. Kite

The purpose of the activity presented in this chapter is fourfold: to (a) have students identify which goal of science (description, prediction, or explanation) is met by a research finding; (b) have students explain that only studies that meet the goal of explanation have a true independent variable and, for those, have students correctly identify the independent and dependent variables; (c) have students explain why research with the scientific goal of prediction cannot demonstrate a causal relationship; and (d) ask students to identify moderator variables for studies with true independent variables.

CONCEPT   Behavioral scientists generate knowledge by asking and answering research questions on a wide range of topics. The goal of both introductory and advanced statistics and research methods courses is to guide students' understanding of the scientific method and how its broad principles apply across research topics. A key concept is that of *variables*: things or concepts that can take on more than one value. Variables can be concrete and directly observable, such as the duration of a brain wave or people's life expectancy, or abstract and not directly observable, such as working memory or visual attention.

To understand the role variables play in the research process, it is important to consider the goal(s) for a particular study. I focus here on three major goals of science: (a) description, (b) prediction, and (c) explanation (see Cozby & Bates, 2015; Whitley & Kite, 2013). Research with a *descriptive* goal uses careful observation or measurement to record events of interest. For example, a researcher might track the rate of post-traumatic stress disorder in Iraqi war veterans or the percentage of female lead actors in feature films over time. When the goal of science is *prediction*, researchers seek to understand whether two variables are systematically related to one another; if so, scientists can make predictions about future behavior. For example, knowing that having a Type A personality is associated with coronary heart disease allows researchers to predict who is more likely to have a heart attack. The third goal of science is *explanation*, or theorizing about why a phenomenon occurs. Scientists typically use the experimental method to establish causality; that is, they manipulate a variable by creating at least two experimental conditions, randomly assign participants to conditions, and strive to ensure that all other variables are held constant. For example, Loftus and Palmer (1974) had participants watch a filmed car crash and estimate the speed of the moving car. The

http://dx.doi.org/10.1037/0000024-016

Activities for Teaching Statistics and Research Methods: A Guide for Psychology Instructors, J. R. Stowell and W. E. Addison (Editors)

wording of the dependent variable was changed so that participants read that the moving car either "smashed" or "contacted" the parked car. Because participants were randomly assigned to conditions, and all other procedures were the same in both conditions, the researchers could conclude that the obtained differences in speed estimates were due to the wording of the question.

Because psychology research focuses on all three goals of science, it is important for instructors to make direct and clear links to the specific scientific goal of a study and the research design. By doing so, they can help students understand when causality can be determined compared with when a pattern is simply a description or when only an association can be established. Because instructors of research methods and statistics often emphasize experimental research, students may not realize that many studies do not have a true independent variable—one in which the experimenter manipulated the variable to create at least two conditions and controlled all other factors. For example, researchers often study whether responses differ by attribute variables such as participant gender or categorization based on a personality variable (e.g., introverts vs. extroverts). Thus, students may not understand that the conclusions researchers draw are linked to the research design used in a particular study.

MATERIALS NEEDED AND INSTRUCTIONS

You should first review the three broad goals of research and the criteria for determining causality. Depending on the students' backgrounds, you may also need to review how to identify independent and dependent variables. Then give students the worksheet (see Appendix 15.1) and work alone to complete Part I of the exercise: categorizing the statements in the handout as description, prediction, or explanation. After most students are finished, they should discuss their answers in groups of four to eight. Next, for each statement, ask one student to explain why her or his group decided on that answer and give students in the other groups a chance to disagree or ask questions. Follow the same procedure for Parts II and III (see Appendix 15.1), which can be completed together or at separate times. (N.B.: It is a good idea to state explicitly to students which four statements to evaluate for each part.)

ASSESSMENT

You can ask students to report the number of statements they categorized correctly working alone. Questions and comments during the discussion phase of the activity are a good source for informally evaluating how well students understand the concepts. You can also provide a new set of research descriptions and collect students' answers to evaluate whether the students are able to generalize their knowledge to new studies. Other effective assessment techniques include the "minute paper" (asking students to report the most important thing they learned or what important question remains unanswered) or the muddiest point (asking students to write down the point that was most confusing; Angelo & Cross, 1993).

DISCUSSION

One issue that arises when using this activity is that it is not always easy to make clear distinctions among descriptions, predictions, and explanations. For example, students can reasonably conclude that Statement 1 ("Eighty percent of people awakened during REM sleep report they were dreaming") is either a description or a prediction. Statement 2 ("Research participants remember words presented at the beginning of a 25-item list better than words presented in the middle of that list . . .") might be seen as prediction because the independent variable (location of word on the list) is not explicitly stated.

This fuzziness has the beneficial effect of encouraging students to ask questions and to focus on their reasons for classifying a statement into a category. For the discussion of Part III, students often correctly note that there are ethical reasons why an independent variable cannot be created; for example, it would not be ethical for researchers to randomly assign people to live with someone they do not like.

Another topic for discussion is that statistical analyses do not, in and of themselves, establish causality. For example, if the results of a t statistic indicate that men exhibit greater aggression than women, this does not mean that being male causes aggression; instead, it shows that aggression is associated with gender. Gender in this example is not a true independent variable but is instead an attribute variable. Instructors can ask advanced students to distinguish between predictor and criterion variables. Advanced students can also identify moderator variables for studies with true independent variables (see Part IV of Appendix 15.1) and can describe the expected pattern of results if those moderators are included. Instructors also can make explicit the connections between the goals of science and research design. For example, most observational and archival studies are descriptive, correlational designs are predictive, and experimental designs show causality. Finally, if your students are conducting a research project on their own or in groups, you can ask them to identify which scientific goal(s) they are addressing and what conclusions they might draw from their results.

REFERENCES

Angelo, T. A., & Cross, K. P. (1993). *Classroom assessment techniques: A handbook for college teachers* (2nd ed.). San Francisco, CA: Jossey-Bass.

Cozby, P. C., & Bates, S. (2015). *Methods in behavioral research* (12th ed.). Boston, MA: McGraw Hill.

Loftus, E. F., & Palmer, J. P. (1974). Reconstruction of automobile destruction: An example of the interaction between language and memory. *Journal of Verbal Learning and Verbal Behavior, 13*, 585–589. http://dx.doi.org/10.1016/S0022-5371(74)80011-3

Whitley, B. E., Jr., & Kite, M. E. (2013). *Principles of research in behavioral science* (3rd ed.). New York, NY: Routledge.

Appendix 15.1

Worksheet and Answer Key

PART I:
DESCRIPTION,
PREDICTION,
AND EXPLANATION

The purpose of this exercise is to help you distinguish among the different goals of science. Your task is to identify whether the statement

A. describes behavior,
B. predicts behavior, or
C. describes the cause of a behavior or explains why a behavior occurs.

Write the letter that corresponds to the goal in the space next to the item. Be prepared to explain why you believe your answer is correct.

1. Eighty percent of people awakened during REM sleep report they were dreaming.
2. Research participants remember words presented at the beginning of a 25-item list better than words presented in the middle of that list because the early words are more likely to have been transferred to long-term memory.
3. In the United States, life expectancy for women is 81.3 years and for men is 76.5 years.
4. People are more aggressive when they cannot be identified than when they can be identified.
5. High school seniors with higher ACT scores have higher college GPAs.
6. The standard deviation on the UCLA Loneliness scale is 1.35.
7. People who like the person they live with have fewer colds than people who do not like the person they live with.
8. Angry faces signal threat, so people are faster at finding an angry face than a happy face in a crowd.
9. When several people witness an emergency, responsibility for offering help is diffused, so the person in need is less likely to receive help, compared to when only one person witnesses the emergency.
10. After training, research participants are able to correctly identify a mockingbird song 52.8% of the time.
11. Memory for digits, as measured by the Digit Span Test, increases with age.
12. Children who show signs of inhibition at 2 months of age are more likely to be shy as teenagers.

PART II:
IDENTIFYING
INDEPENDENT
AND ATTRIBUTE
VARIABLES

Look again at the statements that you classified as "describes the cause of a behavior or explains why a behavior occurs." Write the number of the statement in the blank below and identify the independent variable and dependent variable used in the research. Can a causal relationship be determined? Why or why not?

Statement number	Dependent variable	Independent variable	Can causality be determined? Why or why not?

PART III:
STUDY DESIGN

Look again at the statements that you classified as prediction. For each, explain why a causal relationship cannot be established. Would it be possible to design a study that would show a causal relationship between the variables? If so, describe that study. If not, explain your reasoning.

PART IV:
IDENTIFYING POTENTIAL MODERATING VARIABLES

For each of the statements that you classified as "describes the cause of a behavior or explains why a behavior occurs," think about potential moderating variables, which change or limit the relationship between the independent and dependent variable. Describe how the results would be different if those variables were included.

KEY, PART I

1. a; 2. c; 3. a; 4. c; 5. b; 6. a; 7. b; 8. c; 9. c; 10. a; 11. b; 12. b

KEY, PART II

Statement number	Dependent variable	Independent variable
2	Recall of words	Place of words on list (beginning vs. middle)
4	Aggression	Participant is identified or not
8	Speed of identification	Happy or angry face placed in crowd
9	Likelihood of receiving help	Number of people witnessing emergency (e.g., 10 vs. 1)

Statement number	Dependent variable	Possible moderator variables
2	Recall of words	*Short or long time limit.* The recency effect might be weaker for participants who are given more time to memorize the list (and thus can recall more words overall).
4	Aggression	*Participant gender.* Because men are more aggressive than women, the effect of identifiability on aggression might be stronger for men.
8	Speed of identification	*Race of faces.* Because White people are more likely to associate anger with Black faces than with White faces, they might be fastest at finding angry Black faces.
9	Likelihood of receiving help	*Cost of helping.* People are more likely to provide help if it requires little (e.g., calling 911 on cell phone) versus a great deal of personal cost (e.g., staying at scene until help arrives). Thus, the likelihood of receiving help should be lower in high-cost situations, regardless of the number of bystanders.

16 EVERYTHING IS AWESOME: BUILDING OPERATIONAL DEFINITIONS WITH PLAY-DOH AND LEGOS

Stephanie E. Afful and Karen Wilson

■ ─── ■

In the activity discussed in this chapter students use either Play-Doh or LEGOs to create a three-dimensional structure (e.g., house, boat, car). Pairs of students provide instructions to each other for making the structure with the materials provided. The discussion focuses on the difficulty and accuracy of students' instructions, which serve as their operational definitions.

■ ─── ■

CONCEPT

If you have seen *The LEGO Movie* and heard its theme song, you are most likely aware that "everything is awesome when you're a part of a [research] team." In this chapter we describe an activity involving Play-Doh and LEGOs that is both fun and effective in teaching the concept of operational definitions. The activity promotes active learning, and it stresses the need for precision in operational definitions, which is important given that precision is necessary for replication in psychological studies.

Understanding operational definitions is critical in the research method process, as evidenced by the *APA Guidelines for the Undergraduate Psychology Major Version 2.0* (see Goals 2.1a, 2.4c, 2.4d; American Psychological Association, 2013). *Operational definitions* are empirical definitions that provide meaning based on the "operations" used to produce or measure the construct (Rosenthal & Rosnow, 2008). Students often struggle with the concept of operational definitions, asking such questions as the following:

- How does an operational definition differ from a theoretical definition?
- How do researchers make the intangible somehow tangible?
- How do researchers quantify and measure psychological constructs, and generate or create these definitions?
- When reviewing the literature, where might one find an operational definition within an article?
- How specific or detailed does the operational definition need to be?

It is important to establish a common ground in students' understanding of operational definitions before asking them to formulate their own such definitions.

Textbooks and instructor guides often ask students to generate examples of operational definitions or match the appropriate measurement to the construct (see

http://dx.doi.org/10.1037/0000024-017

Activities for Teaching Statistics and Research Methods: A Guide for Psychology Instructors, J. R. Stowell and W. E. Addison (Editors)

Strohmetz, 2010). For example, Herringer (2000) designed an activity in which students predicted a difference in personality traits between two TV characters (e.g., Captain Kirk and Captain Picard in *Star Trek*). As a class, students created operational definitions of observable behaviors that would represent such traits (e.g., being uptight, being talkative), watched a *Star Trek* episode, and recorded behavior using their operational definitions.

The activity described here requires students to create their own operational definitions. Students use these operational definitions to instruct others in small groups on how to build structures with Play-Doh or LEGOs.

MATERIALS NEEDED

You will need Play-Doh, Ziploc bags, and LEGOs, or their equivalent. You will need to divide the LEGOs into Ziploc bags, making sure each pair of bags has the same pieces. Each pair of students should get one set of identical materials: Ziploc bags with identical LEGOs or identical cans of Play-Doh. The amount of building materials is at your discretion, but to save time we recommend giving a relatively small amount (e.g., two 4-oz cans of Play-Doh or 15 Lego pieces per bag). For larger classes, students can work in groups of three or four.

INSTRUCTIONS

After operational definitions have been reviewed in class, this activity can serve as an effective application of the concept. Allow approximately 15 minutes for the activity and 10 minutes for discussion. In this activity, the three-dimensional structures represent psychological constructs (e.g., memory, happiness, attention), and the instructions to create them represent operational definitions. Specific, step-by-step instructions follow:

1. Divide students into pairs. For larger classes, consider having students work in groups of four, with two pairs working together with an identical set of materials or having two students conduct the activity in front of the class as a fishbowl demonstration. It is also possible for the instructor to participate in a pair or group, but we have found it particularly helpful to circulate during the activity to field any questions students may have. Students should sit back to back so that they cannot see what each other is making.

2. Each pair (or group) should get either two bags of LEGOs or two cans of Play-Doh. Assign each pair of students a structure (e.g., boat, car, house) that they will create (i.e., that can be measured).

3. The first student should build the structure on his or her own, without the other student seeing the creation.

4. After completing the structure, the first student should verbally instruct the second student (while still sitting back to back) in how to make the identical structure. The second student may ask questions of the first student while they make their structures. Make sure there is a pair of students with LEGOs and a pair with Play-Doh making the same structure. For example, as the instructor assigns structures, make sure one pair or group makes a boat with LEGOs while another pair or group makes a boat out of Play-Doh. This will allow for the comparison of the structures made from different building materials after the activity.

5. After the second student has finished creating the structure according to the first student's instructions, have the class retain their creations for class discussion.

6. After all pairs or groups of students have completed their structures, have them compare the original structures with the replications created by the other students, noting the accuracy of the replications and the difficulty of the task when using different building materials.

ASSESSMENT

Assessment can be done using standard multiple-choice exam questions and application-based items for which students are asked to identify the operational definition of a construct. Instructors can also ask students to generate operational definitions of constructs that can be evaluated by other students and/or the instructor.

DISCUSSION

Much of the postactivity discussion involves comparing the same structures made with different materials (LEGOs vs. Play-Doh) and how this relates to different operational definitions, such as when using surveys versus conducting experiments. Although there is a fair amount of variation, the LEGO creations are usually more accurate than those made with Play-Doh because they have a specific size and shape that make it easier to describe how they should be used. Furthermore, the LEGOs are solid, whereas the Play-Doh is malleable and more likely to fall apart. The differences in the types of media used could also lead to a discussion of levels of measurement as you and the students compare the building materials and their relative advantages and disadvantages.

When conducting this activity, be mindful that the second student may attempt to view what the first student has constructed. It is also possible that the first student may try to help the second student gain an advantage by revealing his or her creation. For students to benefit from the activity, it is important to monitor what they are doing and make sure they are not helping each other. We suggest assigning common objects, such as a boat or house, with which everyone is familiar. If students complain that they do not know how to make a specific object, you should instruct them to do the best they can and emphasize that they are not being evaluated on the quality of their structures. A few students have also complained that they do not want to touch the Play-Doh because it would make their hands dirty. In these cases, they can switch to using LEGOs.

When we conduct this activity, students give verbal instructions to their partners. It would also be helpful to invite students to think about how they might write their instructions. It is often harder to convey the instructions in writing, such as in an American Psychological Association–Style research paper. A variation of the activity could include having students convey the steps in writing rather than verbally. You also could divide the class so that some groups give verbal explanations and others give written explanations to their partners, then compare the results.

Operational definitions also aid researchers in replication. During the class discussion, you can ask students to think about the level of detail needed to replicate the psychological constructs (i.e., specify each step, the duration of each step, and how that might translate to levels of measurement). At this point it is also a good idea to emphasize that operational definitions give researchers a common definition for all to use when defining a construct, which facilitates replication. During the activity, students can ask questions of their partners; however, that would not be the case if students were reading a research article and trying to replicate a study given the operational definition provided—hence the need for specific and valid operational definitions in Method sections of research reports.

Students should have fun with this activity. The term *operational definition* should become part of psychology professionals' common language and research toolkit. Introducing students to operational definitions in this stepwise activity should help them in their understanding of the importance of measurement, validity, and precision in the research process.

REFERENCES

American Psychological Association. (2013). *APA guidelines for the undergraduate psychology major version 2.0.* Retrieved from http://www.apa.org/ed/precollege/about/psymajor-guidelines.pdf

Herringer, L. G. (2000). The two captains: The research exercise using "Star Trek." *Teaching of Psychology, 27,* 50–51. http://dx.doi.org/10.1207/S15328023TOP2701_12

Rosenthal, R., & Rosnow, R. L. (2008). *Essentials of behavioral research: Methods and data analysis* (3rd ed.). New York, NY: McGraw Hill.

Strohmetz, D. B. (2010). *Having fun with operational definitions.* Retrieved from http://www.teachpsychscience.org/files/pdf/524201045948PM_1.PDF

17 A DEMONSTRATION OF RANDOM ASSIGNMENT THAT IS GUARANTEED TO WORK (95% OF THE TIME)

Thomas P. Pusateri

Instructors can use the activity introduced in this chapter to demonstrate how random assignment controls for extraneous variables, which strengthens cause-and-effect interpretations of the results. The instructor shuffles and distributes playing cards to students in class to simulate random assignment to an experimental group (e.g., "red card") and control group (e.g., "black card"). He or she then polls students to determine whether the two groups contain roughly equal percentages of students across several demographic categories (e.g., gender, age, location of birth), which helps eliminate alternative causal explanations for the study's results.

CONCEPT

The activity I describe in this chapter (adapted from Pusateri, 2001) is designed to help students understand the rationale for using random assignment to control for extraneous variables that cannot be directly controlled by the experimenter, thereby strengthening the inference of a causal relationship between the independent and dependent variables (i.e., internal validity). May and Hunter (1988) suggested that students often confuse random assignment with random sampling and thus do not grasp the purpose of random assignment in strengthening causal inference. For example, when describing the effects of caffeine consumption on learning, some students may claim that differences between experimental (caffeine) and control (no-caffeine) groups might be due to differences in how well some people tolerate caffeine. However, if random assignment were used appropriately in this study, the distributions of individual differences in tolerance of caffeine should be similar for participants in the experimental and control groups, which would rule out caffeine tolerance as an alternative explanation for the results.

MATERIALS NEEDED

You should purchase enough decks of playing cards so that each student can receive one card. This activity works most effectively in a class of at least 30 students, but it also works reasonably well in classes of 20 to 29 students. If the class has fewer than 20 students, consider using the simulation procedure described later in the Discussion section.

INSTRUCTIONS

One class session before random assignment is discussed, conduct a classroom demonstration of an experiment that produces reliable results, such as a partial replication of Hyde and Jenkins's (1969) study, using the resources available in Appendix 17.1.

http://dx.doi.org/10.1037/0000024-018
Activities for Teaching Statistics and Research Methods: A Guide for Psychology Instructors, J. R. Stowell and W. E. Addison (Editors)

Before reading the word list aloud, randomly distribute the two sets of instructions so that half of the students receive Set A and half receive Set B. Read the words one at a time and ask students to circle Yes or No for each word. After reading the last word, ask the students to write, on the back of the instruction sheet, as many of the 20 words as they can remember. Then read the list and ask students to indicate how many words they correctly recalled in the space on the bottom of the instruction sheet. In general, students who received the instructions in Set B should recall more words than those who received the instructions in Set A. Before collecting the instruction sheets, ask the students to indicate their gender; adapt this question to any characteristic that varies among the students; some suggested characteristics appear later in these instructions. Before the next class, calculate the average number of words recalled by students who received each set of instructions (Set A and Set B), and record the numbers of men and women who received each set of instructions.

In the next class session, use the demonstration above to introduce the logic of experimental design. Ask students to identify the hypothesis, independent and dependent variables, and experimental groups (there are no control groups in this study), and then begin a discussion of extraneous variables and the purpose of random assignment.

When discussing extraneous variables, suggest gender as one possible extraneous variable in this study. For example, assume that the class consists of 30 students, 17 women and 13 men, who collectively received exactly 15 Set A and 15 Set B instruction sheets, one per student. Ask students how the instruction sheets should have been distributed so that gender could not possibly explain the results of the study. At some point, students will indicate that there should be, as close as possible, an even distribution of men and women in each group. In other words, the "best possible case" would be to have nine women and six men receiving one set of instructions, while the other eight women and seven men receive the other set of instructions. Then ask the students what the "worst possible case" would be: In this case, 15 women and no men would receive one set of instructions, while two women and 13 men would receive the other set.

Ask students whether the way the instruction sheets were actually distributed produced a distribution more similar to the best or worst possible case. Students may realize that you did not deliberately pass out the sheets so that they would split evenly between men and women. They may be aware that you distributed the sheets randomly, but they may think that the best and worst possible cases are equally likely to occur. Reveal to them what really happened (e.g., how many men and women actually received each set of instructions). The actual split is typically much closer to the best possible case than to the worst possible case.

To convince them further, shuffle a deck of playing cards containing the same number of cards as students in the class with as equal a number of red and black cards as possible, and then randomly distribute a card to each student. Ask all the women (or men) to stand up, count the number standing, and then ask those holding red cards to sit down. If random assignment produced roughly similar groups, approximately half of those standing should sit down. Explain that random assignment does not always make the groups perfectly equal, but it is much more likely that the split will be closer to the best possible case than the worst possible case (e.g., with 13 women, a 7/6 split is ideal, and you are more likely to get a 7/6, 8/5, or 9/4 split than a 13/0, 12/1, or 11/2 split).

Ask all the students to sit down again and, keeping the same cards, instruct a different group of students to stand and then repeat this process. For example, instruct one of the following five groups to stand: (a) those who were born in the same state (or city) as your institution, (b) those who wear corrective lenses (e.g., glasses, contact lenses, have had LASIK surgery), (c) those who consider themselves outgoing around strangers, (d) those who are the oldest or the only child in their families, (e) student athletes. Vary these questions, add others depending on the composition of the class, or ask students to suggest other characteristics that might affect the results of the study. Each time a group stands, count the total number of students standing. In each case, it is more likely that the students who stand will be split roughly evenly into those holding red cards versus black cards than it will be for everyone standing to have the same color card. Verify this by asking students who are standing and holding a red (or black) card to sit down and then counting the students who remain standing. The main exceptions to this result will occur for questions for which nearly all (or nearly none) of the students will stand up, so it is best to select questions for which approximately half (30%–70%) of the class will stand up. After repeating this process several times, ask students to discuss how many different extraneous variables are roughly equalized among groups via random assignment, which strengthens causal inferences in the interpretation of the study's results.

ASSESSMENT

Include in your assessment test items on the purpose of random assignment in exams or quizzes. Forsyth, Bohling, and Altermatt (1995) provided several research vignettes that can be used to assess students' confidence in causal inferences based on whether or not each research vignette includes random assignment of participants to levels of the independent variable.

DISCUSSION

If the class is small, the chances that an uneven split will occur are increased, which can generate an interesting discussion of the importance of sample size. Consider using the following simulation, which is similar to a demonstration described by Enders, Laurenceau, and Stuetzle (2006). Purchase two decks of pinochle cards and use the 16 kings, 16 queens, and 16 jacks as the 48 "participants" in the simulation. Shuffle these face cards and randomly distribute them between two students. Suggest that jacks represent younger men and kings represent older men in the study, and queens represent women. Have both of the two students count the number of queens they were dealt. It is more likely that the queens will split 8/8, 9/7, or 10/6 than they are to split 16/0, 15/1, or 14/2. Ask the students to count the number of jacks or kings to see if they split closer to 8/8 or 16/0. Then, suggest that red cards and black cards represent some other important dimension, such as political affiliation, and ask students to count how many of the 24 red cards they were dealt. It is much more likely that the red cards will split 12/12 or 13/11 than 24/0 or 23/1.

REFERENCES

Enders, C. K., Laurenceau, J., & Stuetzle, R. (2006). Teaching random assignment: A classroom demonstration using a deck of playing cards. *Teaching of Psychology, 33,* 239–242. http://dx.doi.org/10.1207/s15328023top3304_5

Forsyth, G. A., Bohling, P. H., & Altermatt, T. W. (1995, August). *Developing and assessing students' abilities to interpret research.* Paper presented at the 103rd Annual Convention of the American Psychological Association, New York, NY. Retrieved from http://eric.ed.gov/?id=ED408525

Hyde, T. S., & Jenkins, J. J. (1969). The differential effects of incidental tasks on the organization of recall of a list of highly associated words. *Journal of Experimental Psychology, 82*, 472–481. http://dx.doi.org/10.1037/h0028372

May, R. B., & Hunter, M. A. (1988). Interpreting students' interpretations of research. *Teaching of Psychology, 15*, 156–158. http://dx.doi.org/10.1207/s15328023top1503_17

Pusateri, T. P. (2001, November). A classroom demonstration of random assignment that's guaranteed to work (95% of the time). In D. Johnston (Chair), *Teaching activities and demonstrations II*. Symposium conducted at the 2nd Annual Iowa Teachers of Psychology Workshop, Cedar Rapids, IA.

Appendix 17.1

Instruction Sets

INSTRUCTIONS SET A (Randomly distribute these instructions to half of the class.)

This experiment examines your ability to identify a letter within a word. You will hear a series of words numbered from 1 to 20. When you hear each word, circle "Yes" if the word contains the letter E and "No" if it doesn't contain a letter E.

1. Yes No	5. Yes No	9. Yes No	13. Yes No	17. Yes No
2. Yes No	6. Yes No	10. Yes No	14. Yes No	18. Yes No
3. Yes No	7. Yes No	11. Yes No	15. Yes No	19. Yes No
4. Yes No	8. Yes No	12. Yes No	16. Yes No	20. Yes No

Write the answer to the question that I ask in the following space: _____

What is your gender? _____

INSTRUCTIONS SET B (Randomly distribute these instructions to half of the class.)

This experiment examines your emotional reaction to the sounds of words. You will hear a series of words numbered from 1 to 20. When you hear each word, circle "Yes" if the word sounds pleasant to you and "No" if it doesn't sound pleasant to you.

1. Yes No	5. Yes No	9. Yes No	13. Yes No	17. Yes No
2. Yes No	6. Yes No	10. Yes No	14. Yes No	18. Yes No
3. Yes No	7. Yes No	11. Yes No	15. Yes No	19. Yes No
4. Yes No	8. Yes No	12. Yes No	16. Yes No	20. Yes No

Write the answer to the question that I ask in the following space: _____

What is your gender? _____

WORD LIST (Do not distribute. Read each word aloud, preceding each word with its number. After Word 20, ask students to recall all of the words in two minutes. Read the list again and tell students to "Write the total number of words you recalled in the space on your sheet.")

1. TABLE	5. SALT	9. HIGH	13. DAY	17. PEPPER
2. NIGHT	6. HAPPY	10. ODD	14. GREEN	18. LOW
3. RED	7. NORTH	11. HATE	15. EVEN	19. SAD
4. LOVE	8. MOTHER	12. FATHER	16. CHAIR	20. SOUTH

18 IDENTIFYING CONFOUNDING FACTORS IN PSYCHOLOGY RESEARCH
Chris Jones-Cage

The activity described in this chapter is intended to help students identify possible confounding factors in a psychology experiment and to enhance students' ability to see problem elements when designing their own studies. In the activity, students review brief research descriptions and identify the relationship between variables and confounding factors.

CONCEPT

An awareness of confounding factors is essential to experimental design and manipulations in psychological research. One challenge of planning an experiment is being able to identify and control or eliminate confounding factors that may be related to the independent variables (IVs), thus increasing external validity. Examples of confounding factors include initial differences in the participants (e.g., gender, race, intelligence), differences in how participants are treated, or differences in how participants respond to the experimental conditions. Confounding factors, such as temporal elements (e.g., testing effects, maturation, history effects) and group differences (e.g., mortality and self-selection, differences in responding at the time of testing), also can occur during the experiment. Confounding factors can also be introduced by the sensitivity of the measurement tool, resulting in ceiling or floor effects. The goal of experimental research is to identify differences in the dependent variable (DV) that can be explained by the manipulations of IVs and not confounding factors.

MATERIALS NEEDED AND INSTRUCTIONS

Instructors will need a copy of the handout with sample descriptions for every student (see Appendix 18.1).

Introduce the activity by explaining the definition of *confounding factors*. After distributing the handout with the research descriptions, have students underline the IV(s) and circle the DV. Then assign students to groups of three to four and provide each group with one or more descriptions of a study. Each group should identify at least two possible confounding factors to share with the class because some may be repeated. Instruct each group to report one confounding factor, and then you as the instructor generate a list for each research description. The group should also share with the class how the confounding factor relates to the IV(s) and how they could influence the dependent variable.

The second part of the activity requires students to consider how to correct or control for confounding factors. From the list of confounding factors generated for each

http://dx.doi.org/10.1037/0000024-019
Activities for Teaching Statistics and Research Methods: A Guide for Psychology Instructors, J. R. Stowell and W. E. Addison (Editors)

study, assign one or more factors to each group and ask students to describe how they would "un-confound" the experiment, meaning how they would reduce or eliminate the effect of the confounding factor(s) on the outcome of the experiment. The groups should then present their suggestions to the class.

ASSESSMENT   The same research descriptions can be distributed to the class at a later date. Without referring to the earlier classroom activity, have the students underline the IVs, circle the DV, and place a box around at least one confounding factor for each research description. Corrections can be done in class, with students exchanging their responses and reviewing the answers as a class. Alternatively, their responses can be collected for formal assessment of student learning outcomes.

DISCUSSION   When conducting the activity, the emphasis should be on the following questions:

- What is the relationship between the IV(s) and DV?
- How could confounding factors affect this relationship?
- What can be done to reduce the influence of confounding factors?
- Could the confounding factors have been controlled by the researcher?

The variability of the suggestions during the discussion for un-confounding the experiments should provide opportunities for the class to review elements of experimental design. If one of the requirements for the course is that students develop their own experimental designs, you might consider recommending or requiring that students review their designs for confounding factors.

RESOURCES   Leedy, P. D., & Ormrod, J. E. (2010). *Practical research planning and design.* Upper Saddle River, NJ: Pearson.

Stangor, C. (2015). *Research methods for the behavioral sciences.* Stamford, CT: Cengage.

Weathington, B. L., Cunningham, C. J. L., & Pittenger, D. J. (2010). *Research methods for the behavioral and social sciences.* Hoboken, NJ: Wiley.

Appendix 18.1

Sample Descriptions

1. An industrial/organizational psychologist conducted a field experiment for the Sticky Taffy Company to determine how many defective candies the line workers can identify on the conveyor belt. The researcher tested each worker for 1 hour, varying the speed every 15 minutes, starting with the slowest speed and progressing to the fastest speed. At each speed, the number of defective candies missed was recorded. The researcher reported to the owner of the company that the most effective way for line workers to complete this task was to run the belt at the lowest speed with no more than 50 candies present at a time.

2. A researcher wanted to examine the effects of violent visual content on aggression. He asked 40 men to watch a 10-minute video of a mixed martial arts (MMA) fight. After all participants had viewed the video, they were led to believe that they were also part of another experiment. In this next experiment, participants stated how many blows they would deliver to a man pictured in a fighting ring. The average number of blows reported by the men after viewing the video was five. Using these results, the researcher reported that watching the MMA video increased the level of aggression.

ANSWER KEY *Sample Description 1*

- Independent variables: speed of conveyor belt, number of candies on belt
- Dependent variable: number of defective candies picked out

Possible confounds: experience of the participants, time of day (during the shift) when tested, number of tests run, and varying the presentation of the different speeds with different numbers of candies

Sample Description 2

- Independent variables: MMA presentation, picture of man
- Dependent variable: number of blows

Possible confounds: participants' experience with violence and with MMA, time period between MMA video and picture of man, and the picture itself

19 DEMONSTRATING EXPERIMENTER AND PARTICIPANT BIAS
Caridad F. Brito

In the activity introduced in this chapter, experimenter and participant biases are explained using a class demonstration that highlights both biases at the same time. Postactivity discussion can address some approaches to minimizing these biases in experimental research.

CONCEPT
Although a number of potential biases frequently arise during a psychology study, two of the more common ones are (a) experimenter bias and (b) participant bias. *Experimenter bias* occurs when the researcher is aware of the experimental conditions in which the participants are being tested and, by subtly altering his or her behavior, communicates to the participants the expected outcomes of the study. As a consequence, participants may alter their behavior to conform to the researcher's expectations (Rosenthal, 1966). Experimenter bias is applicable to various types of research methods, but it is most commonly seen in experimental designs in which the researcher behaves differently toward the participants in one group compared with the participants in other groups. This type of bias is generally not purposeful behavior; in fact, in most instances the researcher is unaware of any differential treatment of participants.

The subtle communication of a researcher's expectations about the outcome of a study may create participant bias. Broadly defined, *participant bias* occurs when participants' anticipations or thoughts about a study influence their responses and, thus, the study's results. In experimental methodology, participant bias is most often seen in the form of *demand characteristics*: cues to which participants attend during the study that may subsequently lead them to alter their behavior in ways that they believe correspond to, or in some cases contradict, what the researcher expects to find (de Munter, 2005). Demand characteristics may arise from the participants' own motivations or from elements within the experimental setting itself, including interactions with the researcher, prior experiences as a research participant, or even rumors about the purpose of the study. The activity described herein deliberately incorporates both biases and helps students learn to distinguish between them.

http://dx.doi.org/10.1037/0000024-020
Activities for Teaching Statistics and Research Methods: A Guide for Psychology Instructors, J. R. Stowell and W. E. Addison (Editors)

Simple images of concrete objects (e.g., cars, animals, kitchen items) or line drawings work well, although other stimuli could be used. There are many images available online for free, but instructors should be mindful of any copyright restrictions.[1] Only 10 colored images are needed. You should select images of simple, clearly identifiable objects and add them to a Word document or PowerPoint slide, with one image to a page or slide. Then create a copy of each of the 10 images. Using Office's "Picture Tools," change the color saturation of each copied image to remove its color, making it gray scale. Of course, there are other techniques for generating the images (e.g., taking your own photographs of household objects, drawing simple figures). Ultimately, you want to have 10 colored images and the same 10 images in gray scale. You should number the images and present them on a screen to all of the students in the class at the same time. Students will also need a writing instrument and a sheet of paper on which to record their ratings of the images.

You should formulate a hypothesis for the demonstration stating that the color and grayscale images are expected to differ on some feature (e.g., colored images are more interesting, funnier, or more attention grabbing than grayscale images). Do not share your hypothesis with the students. Tell the students that you are going to show them 20 images, one at a time, and they should rate how interesting or funny or attention grabbing (this measure should be adjusted depending on the images used) they find that image. You can use any scale you prefer—1 to 5 works well, with 1 being the lowest and 5 being the highest on the feature being rated. Then, randomly present one image at a time to the class; for each image, students record their ratings on the sheet of paper.

To create experimenter bias, while subtly developing participant bias, you should make offhand comments and sounds every few images and show excitement through body movements and facial expressions. These comments, body movements, and facial expressions are designed to influences students' answers to support your hypothesis. For example, for colored images, you might say "Wow," "Oh," "Interesting," while nodding your head up and down, raising your eyebrows, and opening your eyes wide; for grayscale images, you could grunt, scrunch your shoulders, and/or frown. If possible, it is a good idea to move around the room so that all of the students have a chance to see your reactions to the images.

After the students have rated the images you can begin a discussion by asking the students to guess what the hypothesis was. Typically, they will be able to identify the hypothesis that colored images are more "interesting" (or whatever feature you had them rate) than grayscale images. Next, ask them what it was about the "study" that helped them identify the hypothesis. You will find that most of the students will have noticed that you behaved differently when colored and grayscale images were displayed. Ask them to specify what it was that they observed you do, or what it was

---

[1]If you do a Google image search, select "Search Tools," then "Usage Rights," then select one of the "Reuse" options. "Labeled for Reuse" is a good choice because you can save and copy the image. If you do a Bing image search, under "License" you can select from among several copyright options. You can search for simple images by adding the term *clipart* to the search phrase.

that you did differently. You can then explain the concept of experimenter bias. In general, this is the bias that is detected more easily by the students. You should also ask them if they noticed any changes in their own behavior or ratings; for example, you can ask them what they were thinking when they were noticing your behaviors in response to the images and whether their ratings conformed to your expectations. These questions allow you to segue into a discussion of demand characteristics and participant bias.

After the discussion, collect the students' sheets and do a quick calculation of the mean ratings for each image. The biases will, however, be more clearly demonstrated by a show of hands. For instance, you can present the images again, one at a time, and ask students to raise their hands if they had rated the image less than 3, and how many rated it higher than 3, using a 5-point scale. You can keep a rough tally on the board for each image. Identify which of the numbered images were the colored images and which were the grayscale images. In general, you will find that the colored images have higher ratings than the grayscale images. Depending on the number of students in the class and the time available, a more formal analysis could be done. You could create two columns, one for the color version and one for the noncolor version of each image, and then ask each student for his or her score for each version of the image. Then, conduct a paired-samples t test to compare the means. The analysis can be done as part of the class activity, or you can do the analysis on your own and share the results with the students at the next class.

ASSESSMENT

Student learning can be assessed in a number of ways. One approach is to give an essay-type quiz asking the student to define participant and experimenter bias and to compare and contrast them. You can also ask an applied question that requires students to create their own examples of each type of bias in the context of a specific study. Other types of assessments include providing the students with a few descriptions of studies in which participant and/or experimenter bias have occurred and asking them to identify whether there is bias and, if so, what type.

DISCUSSION

There are various approaches to minimizing the impact of demand characteristics on research results. An obvious approach is to not reveal to the participants the hypothesis being tested, instead focusing on the behaviors in which the participants will have to engage (rating images). This approach might be considered deception by omission (i.e., passive deception) whereby the researcher does not directly mislead or lie to the participants but instead simply does not reveal to the participants all of the details of the research being conducted. Of course, debriefing after any kind of experimental manipulation is highly recommended and usually required if participants were misled in any way (American Psychological Association, 2010).

Another approach a researcher might use to reduce both demand characteristics and experimenter bias is by being as neutral as possible—the opposite of what is demonstrated in this activity (Podsakoff, MacKenzie, Lee, & Podsakoff, 2003). Because being neutral is not easily achieved given how subtle these biases can be, the use of double-blind designs may be appropriate when using placebo or control groups (i.e., the researcher who is interacting directly with participants does not know to which condition of the study a participant is assigned or how the participant should respond; similarly, the participants are unaware of the treatment condition to which they have been assigned). Although it is probably impossible to completely eliminate these biases, minimizing their impact will make the conclusions drawn from an experiment more valid.

ADDITIONAL RESOURCE

Nichols, A. L., & Maner, J. K. (2008). The good-subject effect: Investigating participant demand characteristics. *The Journal of General Psychology, 135,* 151–166. http://dx.doi.org/10.3200/GENP.135.2.151-166

REFERENCES

American Psychological Association. (2010). *Ethical principles of psychologists and code of conduct (2002, amended June 1, 2010)*. Retrieved from http://www.apa.org/ethics/code/principles.pdf

de Munter, A. (2005). Demand characteristics. In B. S. Everitt & D. Howell (Eds.), *Encyclopedia of statistics in behavioral science* (pp. 477–478). Hoboken, NJ: Wiley. http://dx.doi.org/10.1002/0470013192.bsa166

Podsakoff, P. M., MacKenzie, S. B., Lee, J. Y., & Podsakoff, N. P. (2003). Common method biases in behavioral research: A critical review of the literature and recommended remedies. *Journal of Applied Psychology, 88,* 879–903. http://dx.doi.org/10.1037/0021-9010.88.5.879

Rosenthal, R. (1966). *Experimenter effects in behavioral research*. New York, NY: Appleton-Century-Crofts.

20 THE MOST UNETHICAL RESEARCHER: AN ACTIVITY FOR DEMONSTRATING RESEARCH ETHICS IN PSYCHOLOGY

Sue Frantz

Listening to a lecture on research ethics can be boring for students. "The Most Unethical Researcher" activity, described in this chapter, gives students an opportunity to learn what it means to conduct ethical research by contrasting it with unethical research. At the conclusion of this activity students will be well-versed in the American Psychological Association's research and publication ethics standards.

CONCEPT

The American Psychological Association's (APA's) *Guidelines for the Undergraduate Psychology Major Version 2.0* (2013) identifies five goals for the major. The first outcome for "Goal 3: Ethical and Social Responsibility," is "Apply ethical standards to evaluate psychological science and practice." The first two of the foundation indicators for this outcome are the following:

1. Describe key regulations in the APA Ethics Code for protection of human or nonhuman research participants.
2. Identify obvious violations of ethical standards in psychological contexts. (APA, 2013, p. 26)

The activity[1] described in this chapter will give students experience working with the APA's *Ethical Principles of Psychologists and Code of Conduct* (2010, hereinafter referred to as the *Ethics Code*) in a low-pressure, classroom environment. The Ethics Code comprises 10 Standards. The one most relevant to researchers is Standard 8: Research and Publication.

MATERIALS NEEDED

Each student will need a paper copy or digital access to the 18-page Ethics Code. In addition, every two or three students will need a copy of "The Most Unethical Researcher" handout, see Appendix 20.1.

INSTRUCTIONS

After introducing students to the basic elements of the Ethics Code, including the General Principles and the 10 Standards, you should inform students that they will be envisioning an unethical psychology scientist who has managed to violate all

---

[1]I wrote a simplified version of this activity that first appeared in an APA Board of Educational Affairs working group report (McCarthy, Frantz, Gurung, McEntarffer, & Ocampo, 2016).

http://dx.doi.org/10.1037/0000024-021

Activities for Teaching Statistics and Research Methods: A Guide for Psychology Instructors, J. R. Stowell and W. E. Addison (Editors)

15 elements of Standard 8. Give students about 10 minutes to read through the two pages of Standard 8. Once they have finished reading, ask students to work in pairs or groups of three to create an unethical researcher using "The Most Unethical Researcher" handout.

Instruct students to spend a few minutes picturing a hypothetical person at your institution, including, but not limited to, the name and the kind of research this person does. Once they have someone in mind and they have written their description on the handout, students will briefly summarize each ethical element and provide an example of each of the researcher's ethical violations. The example should illustrate how violating the ethical principle could lead to serious problems. A weak example would be that their imaginary researcher did not get institutional review board approval before conducting the study. A better example would be that the researcher chose to conduct an experiment to evaluate what makes students more stressed, not having caffeine or having too much caffeine, without having the study approved by their institution.

As students work together, circulate around the room answering any questions they may have about the Ethics Code and ensuring that they are creating detailed examples. When the groups appear to be done, start with the first element of Standard 8.01P: Institutional Approval. Read the element aloud to the class and then ask a volunteer or two to share their group's example of a violation of that element. After each example, ask the class how the researcher could have avoided violating that ethical principle and why it is important to adhere to that ethical principle. Repeat this process for the remaining 14 elements. At the end of class, collect the completed handouts.

ASSESSMENT   For the first outcome, "Describe key regulations in the APA Ethics Code for protection of human or nonhuman research participants," review the responses in the "Short Summary" column of the table in Appendix 20.1. If using a grading rubric, for each response you may choose to award 2 points for a good summary that addressed the major elements of the ethical principle, 1 point for an incomplete summary (addressed some but not all of the major elements), and 0 points for an incorrect or missing summary. For the second outcome, "Identify obvious violations of ethical standards in psychological contexts," review the responses in the "Ethics Code Violation Examples" column. For each response, you can award 2 points for a specific example that addressed the major elements of the ethical principle, 1 point for an accurate but nonspecific example, and 0 points for an incorrect or missing example.

You should enter scores for each group into a spreadsheet for easy comparison. Are there some ethical principles students had a more difficult time summarizing or creating an example for than others? If so, consider addressing those during the next class session.

DISCUSSION   Because students are encouraged to envision over-the-top, bad behavior, do not be surprised if you hear much laughter in the room as students create their unethical researcher. Pay particular attention to groups that are not laughing. Those groups may not be creating sufficiently specific examples.

At the conclusion of the activity, ask students what they would do if this person were a fellow student or a faculty member at your institution. Be prepared to tell students what your institution's policies are regarding the reporting of ethics violations.

If class time allows, collect the completed handouts and choose some examples to read aloud. You may also invite students to vote on the most egregious examples. Can students identify which ethical principle was violated? Alternatively, either through your course management system or as a handout, create a matching question using the examples your students created. On one side of the matching question, list 15 examples taken from the ones your students created, one for each ethical principle. On the other side, list the 15 ethical principles. The matching question could be used as study practice, as an out-of-class assignment, or as an examination question.

If you would like your students to explore research ethics in greater depth, APA's (2016) *Responsible Conduct of Research* includes topics such as collaborative science, laboratory animal welfare, and mentoring. Each topic page includes a short summary and a set of supporting documents. Each student group could review the supporting documents for a particular topic with the goal of providing that information to their fellow students, either in writing, as a poster, or as an in-class presentation.

REFERENCES

American Psychological Association. (2010). *Ethical principles of psychologists and code of conduct (2002, amended June 1, 2010)*. Retrieved from http://www.apa.org/ethics/code/principles.pdf

American Psychological Association. (2013, August). *APA guidelines for the undergraduate psychology major: Version 2.0*. Retrieved from http://www.apa.org/ed/precollege/about/psymajor-guidelines.pdf

American Psychological Association. (2016). *Responsible conduct of research*. Retrieved from http://www.apa.org/research/responsible/index.aspx

McCarthy, M. A., Frantz, S., Gurung, R. A. R., McEntarffer, R., & Ocampo, C. (2016). *Assessment of outcomes of the introductory course in psychology*. Working group report presented to the American Psychological Association Board of Educational Affairs, Washington, DC.

Appendix 20.1

The Most Unethical Researcher

Names of students in group:

   Instructions: Your task is to imagine a researcher who has managed to violate all elements of Standard 8, Research and Publication, of the American Psychological Association's *Ethical Principles of Psychologists and Code of Conduct* (hereafter *Ethics Code*).

   Researcher description: Provide a brief description of your imaginary researcher, such as name, place of employment, and a little bit about the kind of research this person does.

   Ethics code violation examples: In the "Short Summary" column of the following table, provide a summary of five words or fewer summary that captures the essence of each ethical element and then provide a short narrative in the far right column that describes how your researcher managed to violate that element of the code. A good example of violating Standard 8.01, Institutional Approval, would be, "The researcher, in submitting the institutional review board approval application, wrote that participants would not be subject to any stress, when in fact the researcher planned to keep participants in sensory deprivation chambers for 9 days," whereas a poor example would be "The researcher lied to the institutional review board."

Standard	Short summary	Ethics code violation examples
8.01: Institutional Approval		
8.02: Informed Consent to Research		
8.03: Informed Consent for Recording Voices and Images in Research		
8.04: Client/Patient, Student, and Subordinate Research Participants		
8.05: Dispensing With Informed Consent for Research		
8.06: Offering Inducements for Research Participation		
8.07: Deception in Research		
8.08: Debriefing		
8.09: Humane Care and Use of Animals in Research		
8.10: Reporting Research Results		
8.11: Plagiarism		

Standard	Short summary	Ethics code violation examples
8.12: Publication Credit		
8.13: Duplicate Publication of Data		
8.14: Sharing Research Data for Verification		
8.15: Reviewers		

21 THE ETHICS OF BEHAVIORAL RESEARCH USING ANIMALS: A CLASSROOM EXERCISE

Harold Herzog

The use of nonhuman animals in behavioral and biomedical research raises complex ethical issues. In the activity described in this chapter, groups of students serve as members of hypothetical university animal ethics committees. The groups evaluate the potential costs and benefits of psychology experiments involving animals and decide whether to approve or reject scenarios that are based on actual studies.

CONCEPT   Psychology research with animals has long been controversial. In 1907, John B. Watson, the founder of behaviorism, was attacked by the press for a series of experiments in which rats were blinded and deafened (Dewsbury, 1990). More recently, in his book *Animal Liberation*, often referred to as the "bible of the animal rights movement," the philosopher Peter Singer (1975) targeted classic behavioral experiments with animals that he deemed cruel and unnecessary. These included Harry Harlow's studies on maternal deprivation in monkeys and Martin Seligman's experiments on the use of electric shock to create learned helplessness in dogs.

The debate over the use of animals in research continues today, and public opinion is deeply divided on the issue. A recent survey conducted by the Pew Research Center (2015) found that 47% of American adults approved of the use of animals in scientific research, and 50% disapproved.

The primary legislation governing animal research in the United States is the Laboratory Animal Welfare Act of 1966 (Pub. L. No. 89-544; hereafter AWA). Amendments to the AWA enacted in 1985 required that colleges and universities in which animals are used for research or educational purposes establish Institutional Animal Care and Use Committees (IACUCs). These committees conduct regular inspections of animal research facilities and review experiments involving animal subjects. Note that rats, mice, and birds—the species that make up over 95% of animals used for research in the United States—are exempt from coverage under the AWA. All vertebrate species, however, are covered under the Public Health Service Policy on the Humane Care and Use of Laboratory Animals (Office of Laboratory Animal Welfare, U.S. Department of Health and Human Services, 2015). As a result, in spite of the exemption under the AWA, nearly all university IACUCS do review experiments involving rats, mice, and birds.

The activity described herein is designed to facilitate classroom discussions of ethical issues associated with the use of animals in research. Small groups of students evaluate the costs and benefits of animal research proposals and make decisions about

http://dx.doi.org/10.1037/0000024-022
Activities for Teaching Statistics and Research Methods: A Guide for Psychology Instructors, J. R. Stowell and W. E. Addison (Editors)
Copyright © 2017 by the American Psychological Association. All rights reserved.

whether the research should be permitted. In an optional second activity, a brief scale is used to assess student attitudes about the use of animals in research. This can lead to discussions of issues such as the origins of gender differences in beliefs about the treatment of other species.

MATERIALS
NEEDED

Appendix 21.1 includes two sample proposals. Additional proposals can be found at http://files.harpercollins.com/OMM/SomeWeLoveResources.pdf. Appendix 21.2 includes the 10-item version of the Animal Attitudes Scale (AAS-10; Herzog, Grayson, & McCord, 2015), a measure of individual differences in attitudes toward the treatment of animals.

INSTRUCTIONS

Divide the class into groups of five to seven students and select a group leader, who should be given a copy of the animal research proposal. The leader reads the proposal to the group members and leads a discussion of the proposal. You should encourage the groups to discuss the pros and cons of the study and try to reach a consensus rather than taking a quick vote on whether to approve the research. Among the issues the students should consider are the following: What will be gained by the experiment? Will the research potentially lead to more effective treatments for human disorders? Will it answer an important scientific question? The students should also consider the degree of harm or pain and suffering for the animals in the study.

Depending on the time available, give each group one or two proposals to review. This process usually takes 15 to 20 minutes. Once the groups have made their decisions, ask the group leaders to come to the front of the room and briefly describe the study to the rest of the class. They should report their group's decision to approve or reject the research proposal and their reasons for making that decision. At this point, other students in the class can weigh in on the proposed research. Often, groups will make different decisions for the same proposals. This is to be expected. A study of actual IACUCS found high levels of disagreement in decisions between committees reviewing the same proposals (Plous & Herzog, 2001).

Although not necessary, I have students read and discuss "Human Morality and Animal Research" (Herzog, 1993) before participating in the exercise. Although the article was written in 1993, the issues it raises related to moral consistency and the treatment of animals have not changed.

The research proposals are based on actual experiments. You should, however, feel free to modify the proposals or to write your own that address other ethical issues, such as housing, training of laboratory personnel, and veterinary care. A variant of this exercise can be used to facilitate class discussions of ethical issues associated with the use of human subjects. In this case, the students serve as members of hypothetical institutional review boards.

ASSESSMENT

You may want to administer to your students the AAS-10 (see Appendix 21.2) before and/or after discussion of the animal research proposals, to assess general changes in attitudes about animal research. This is a brief form of a scale that has been widely used to assess attitudes toward the use of other species. The 10-item version has excellent psychometric properties (Herzog et al., 2015), and it can be used to facilitate discussions of an array of topics related to the psychological aspects of relationships with other species. For example, women will usually have, on average, higher AAS-10 scores than men, indicating greater concern for animal welfare.

The use of animals in biomedical and behavioral research is among the thorniest of ethical issues associated with the treatment of other species. Group members often initially disagree about approval/rejection decisions. Sometimes group members come to a consensus decision after discussing the costs and benefits of the proposals. In other cases, however, they remain divided. When this happens, I encourage dissenting students to inform the class why they disagreed with their fellow group members. Some students will have strong feelings about the use of animals in science. In my experience, however, they tend to be respectful of each other's views when discussing the merits of specific proposals during this exercise.

ADDITIONAL RESOURCE

Reiss, D., & Marino, L. (2001). Mirror self-recognition in the bottlenose dolphin: A case of cognitive convergence. *Proceedings of the National Academy of Sciences of the United States of America, 98,* 5937–5942. http://dx.doi.org/10.1073/pnas.101086398

REFERENCES

Dewsbury, D. A. (1990). Early interactions between animal psychologists and animal activists and the founding of the APA Committee on Precautions in Animal Experimentation. *American Psychologist, 45,* 315–327. http://dx.doi.org/10.1037/0003-066X.45.3.315

Herzog, H. (1993). Human morality and animal research: Confessions and quandaries. *The American Scholar, 62,* 337–349.

Herzog, H. (2010). *Some we love, some we hate, some we eat: Why it's so hard to think straight about animals.* New York, NY: Harper.

Herzog, H., Grayson, S., & McCord, D. (2015). Brief measures of the Animal Attitude Scale. *Anthrozoös, 28,* 145–152. http://dx.doi.org/10.2752/089279315X14129350721894

Laboratory Animal Welfare Act of 1966, Pub. L. No. 89-544, 7 U.S.C. § 2131 *et seq.*

Langford, D. J., Crager, S. E., Shehzad, Z., Smith, S. B., Sotocinal, S. G., Levenstadt, J. S., . . . Mogil, J. S. (2006, July 27). Social modulation of pain as evidence for empathy in mice. *Science, 312,* 1967–1970. http://dx.doi.org/10.1126/science.1128322

Office of Laboratory Animal Welfare, U.S. Department of Health and Human Services. (2015). *Public health service policy on the humane care and use of laboratory animals.* Washington, DC: National Institutes of Health. Retrieved from http://grants.nih.gov/grants/olaw/references/phspolicylabanimals.pdf

Pew Research Center. (2015, July 1). *Americans, politics and science issues.* Retrieved from http://www.pewinternet.org/files/2015/07/2015-07-01_science-and-politics_FINAL.pdf

Plous, S., & Herzog, H. (2001, July 27). Animal research: Reliability of protocol reviews for animal research. *Science, 293,* 608–609. http://dx.doi.org/10.1126/science.1061621

Singer, P. (1975). *Animal liberation.* New York, NY: Avon Books.

Appendix 21.1

Animal Research Proposals

Your group is the Institutional Animal Care and Use Committee (IACUC) for your university. It is the committee's responsibility to evaluate and either approve or reject research proposals submitted by faculty members who want to use animals for their studies. The proposals briefly describe the experiments, including their goals, potential benefits, and the costs of the research in terms of possible harm or discomfort to the animals. Your group must either approve or deny permission for the experiment to be conducted. Note that it is not your job to critique technical aspects of the projects, such as the experimental design. You should make your decision on the basis of the information given in the proposal.

PROPOSAL 1: THE EFFECTS OF MATERNAL DEPRIVATION ON BRAIN DEVELOPMENT IN MONKEYS

An estimated 3 million adolescents suffer from major depression each year. Dr. Kaleen uses a nonhuman primate model to study the developmental neurobiology of this disorder. He is requesting permission from your committee to conduct a study of the effects of maternal deprivation on brain development in young monkeys. Twenty baby rhesus monkeys will be taken from their mothers at birth. For the first month of their lives they will be raised in isolation. For the next 11 months they will be housed with another baby monkey. Every 3 weeks, the animals will be subjected to a short-term stressor, such as exposure to a harmless snake or a noisy novel toy. Immediately following each of these trials, small amounts of blood will be drawn to assess the animals' levels of cortisol, a stress hormone. In addition, every 4 months the monkeys will be anesthetized and given an MRI (a type of brain scan). This will allow Dr. Kaleen to study the development of areas of the primate brain involved in depression. The control group will consist of 20 monkeys who will stay with their mother for the entire year after they are born. They will also be given the stress tests, blood tests, and MRIs.

After 12 months, animals in both the experimental and the control groups will be painlessly euthanized. Their brains will then subjected to detailed anatomical and neurochemical analyses.

Dr. Kaleen notes that his research is based on studies of the effects of isolation in monkeys by Harry Harlow, which showed that maternal deprivation can produce long-term depression in animals. But, unlike Harlow's studies, the baby monkeys will be only be completely isolated for 1 month. Kaleen notes that little is known about how maternal deprivation affects brain development. He argues that the results of this study will help lead to the development of new types of treatments for depression in teenagers and young adults.

Approve or reject the study? Give the reasons for your decision.

(Note to instructor: This scenario is based on a protocol that was initially rejected but subsequently approved by the IACUC of a major university with modification [a reduction in the length of social isolation]).

Dr. Jones studies pain using nonhuman animals. In his laboratory, pain is induced in mice by injecting the animals in their stomachs with a solution of acetic acid. This procedure produces a sequence of behaviors called the "writhing response." In a previous study, Jones showed that mice are capable of experiencing a type of empathy when in pain: When tested in pairs, mice show higher levels of pain responses than when tested alone—but only if the second mouse is a former cage mate rather than a stranger.

Dr. Jones is requesting permission from your committee to conduct an experiment that will show whether the mice communicate their distress to former cage mates through vision, sound, or smell. The study will involve 150 mice. Half of the mice will be "stimulus" animals. They will simply be injected with the acetic acid solution. The other 75 mice will be "observers." These animals will also be injected in the stomach with the acid solution, but, in addition, they will be deprived of one of their senses. The first group of animals will be blocked from seeing the stimulus animal by having an opaque barrier placed between them and their former cage mate. The second group of 25 mice will first be permanently deafened by having a chemical injected into their ears every day for 2 weeks. The third group will be permanently deprived of their ability to smell by having a toxin inserted into their noses that destroys odor-detecting nerve cells.

The mice will be given the writhing test in pairs to see which group of the observer animals shows empathy responses in the paired writhing test. Dr. Jones says the design of the experiment will allow him to determine whether the stimulus animals are communicating their pain by visual signals, high-frequency sounds, or chemical signals.

Approve or reject: What is the reason for your decision?

(Note to instructor. This scenario is based on series of studies that provided the first evidence that nonprimate animals experience empathy [Langford et al., 2006]. The experimenters found that mice deprived of their sense of hearing or smell still showed the empathy response, whereas the response was eliminated in mice who could not see other animals in pain. The study involved many hundreds of animals that were subjected to pain and/or permanent destruction of their sensory capacities. Given the degree of suffering involved in the research, it is ironic that it was lauded by some animal activists who argued that it demonstrated parallels between the emotional responses between mice and humans. [See Herzog (2010, Chapter 8) for a discussion of the paradox of using the results of studies of animal behavior to argue that animal research should be eliminated.])

Appendix 21.2

The Animal Attitudes Scale (10-Item Version)

The AAS-10 consists of all the items below. Higher scores indicate more concern for animal welfare. The numbers of points assigned to the response items are in parentheses. Starred items (*) are reverse scored. For psychometrics of the scale, see Herzog, Grayson, and McCord (2015).

Instructions: Listed below are a series of statements regarding the use of animals. Circle the letters that indicate the extent to which you agree or disagree with the statement:

SA = Strongly Agree (5)
A = Agree (4)
U = Undecided (3)
D = Disagree (2)
SD = Strongly Disagree (1)

1. It is morally wrong to hunt wild animals just for sport.

   SA A U D SD

2. I do not think that there is anything wrong with using animals in medical research.*

   SA A U D SD

3. I think it is perfectly acceptable for cattle and hogs to be raised for human consumption.*

   SA A U D SD

4. Basically, humans have the right to use animals as we see fit.*

   SA A U D SD

5. The slaughter of whales and dolphins should be immediately stopped even it means some people will be put out of work.

   SA A U D SD

6. I sometimes get upset when I see wild animals in cages at zoos.

   SA A U D SD

7. Breeding animals for their skins is a legitimate use of animals.*

   SA A U D SD

8. Some aspects of biology can only be learned through dissecting preserved animals such as cats. *

SA A U D SD

9. It is unethical to breed purebred dogs for pets when millions of dogs are killed in animal shelters each year.

SA A U D SD

10. The use of animals such as rabbits for testing the safety of cosmetics and household products is unnecessary and should be stopped.

SA A U D SD

Note. From "Brief Measures of the Animal Attitude Scale," by H. Herzog, S. Grayson, and D. McCord, 2015, *Anthrozoös, 28*, p. 152. Copyright 2015 by Routledge. Reprinted with permission.

22 DEMONSTRATING INTEROBSERVER RELIABILITY IN NATURALISTIC SETTINGS

Janie H. Wilson and Shauna W. Joye

The hands-on exercise described in this chapter focuses on interobserver reliability. Using naturalistic observation, students work in teams to collect data in a public location. Students learn the importance of discussion and training to enhance interobserver reliability.

CONCEPT

Interobserver, or interrater, reliability addresses the ability of more than one researcher to collect similar data on variables of interest. Assessment of interobserver reliability is particularly important when observing subjective behaviors, such as friendliness among children at recess. According to Salkind (2010), researchers can assess interobserver reliability with Pearson's r, with an r value of .70 or higher indicating acceptable consistency between researchers.

MATERIALS NEEDED

This activity requires a handout (see Appendix 22.1) that explains what students need to do and contains tables to be completed by them. If required by the instructor, students will also need access to data-analysis software such as SPSS. As an alternative, students can calculate interobserver reliability by hand.

INSTRUCTIONS

Before conducting this activity, instructors who have access to an institutional review board should gain permission for students to collect data. At the beginning of class, provide the handout to each student and randomly assign students to teams of two. When a class contains an odd number of students, allow one team of three, requesting that each student complete all parts of the activity.

Teams go to a public campus location, such as a dining hall, food court, student union, or so on, and unobtrusively observe people at the location. Without discussing the measure, students observe peoples and rate them on creativity using a rating scale that ranges from 1 (*not creative at all*) to 10 (*very creative*). Note that these instructions are vague and can be interpreted in a variety of ways. One of the main goals of the activity is for students to discover that they will need to operationalize the variable of creativity to obtain good interobserver reliability.

After collecting data from 12 people, students review their ratings and discuss reasons for variability between observers. Next, they observe five more people, discussing reasons why they assigned their specific ratings immediately after each observation. Students attempt to define exactly what they are measuring (creativity) and how they

http://dx.doi.org/10.1037/0000024-023

Activities for Teaching Statistics and Research Methods: A Guide for Psychology Instructors, J. R. Stowell and W. E. Addison (Editors)

are measuring it. The goal is to operationally define *creativity* and reach a consensus on assessment. Next, raters observe 12 more people without further discussion of ratings.

For each of the 12-person data sets, students calculate Pearson's *r* for the pairs of scores from the two raters. The second data set of 12 people should yield a higher Pearson's *r* than the first data set of 12.

ASSESSMENT   Assessment questions are found at the end of Appendix 22.1. Items require students to consider the importance of operational definitions, researcher training, and interobserver reliability.

DISCUSSION   While completing this naturalistic observation activity students learn the importance of training for interobserver reliability. Depending on the requirements you give them, students may also learn how to analyze their data using Pearson's *r* as a reliability measure. Assessment in the form of open-ended questions encourages thought and enriches class discussion.

Additional teaching possibilities include the following options:

- Students can calculate Pearson's *r* by hand, depending on whether you deem it necessary.
- Students can analyze the data using a computer program (e.g., SPSS) and print the output page as part of the activity assessment.
- Students can write up the activity in the form of American Psychological Association–Style Method and Results sections.
- A class discussion could be held that focuses on the ethics of observational research, including concerns about judging a construct such as creativity by simply observing people.
- The activity could include an explanation of Likert and other types of scales as well as alternative ways to assess creativity (operationalize variables).
- Instructors could also give examples of their own experiences with interobserver reliability.

REFERENCE   Salkind, N. J. (Ed.). (2010). *Encyclopedia of research design* (Vol. 1). Thousand Oaks, CA: Sage. http://dx.doi.org/10.4135/9781412961288

Appendix 22.1

Activity Handout

The purpose of this activity is to learn about interobserver reliability in a naturalistic setting. In teams of two, you will collect data on campus using naturalistic observation. With your partner, use the assigned time period to observe individuals in a public place, such as a dining hall, food court, student union, etc. Be careful not to interact with anyone or make it obvious that you are recording others' behavior, which might make people feel uncomfortable.

Without discussing this assignment with your teammate and using your *own* handout, each of you should rate the creativity of the same 12 people on a scale from 1 to 10 (1 = *not creative at all*, 10 = *very creative*). Write your ratings in the table below under the column titled "Observation Time 1." The only requirement is that you and your teammate *must* evaluate the same person each time.

Next, discuss with your teammate how to rate creativity until you are in agreement of how to assess it exactly the same way. Then rate one new individual together, discussing why you would choose a specific rating. Continue to discuss ratings for the next four people observed. *The goal is to operationally define creativity in such a way that both of you will arrive at similar creativity ratings for each person.* You do not need to record these five training ratings; you just need to agree on them.

After you have discussed ratings for five people, rate 12 new people individually and enter your ratings in the column labeled "Observation Time 2" in the second table Again, you both must indicate who is being evaluated so you can rate the same person each time. Otherwise, do not interact with your teammate. After you have entered your own ratings, enter your partner's ratings in the columns next to your ratings.

Observation Time 1				Observation Time 2		
Person observed	Your ratings	Your partner's ratings		Person observed	Your ratings	Your partner's ratings
1				1		
2				2		
3				3		
4				4		
5				5		
6				6		

Observation Time 1		
Person observed	Your ratings	Your partner's ratings
7		
8		
9		
10		
11		
12		

Observation Time 2		
Person observed	Your ratings	Your partner's ratings
7		
8		
9		
10		
11		
12		

Analyze both sets of observational data separately with Pearson's r to determine how well you and your teammate's ratings agree.

Time 1 Pearson's r: _____   Time 2 Pearson's r: _____

Among other things, Pearson's r can be used to assess interobserver reliability, with higher values representing better reliability. Because you and your teammate assessed the same people, your ratings should be related, with higher values by you associated with higher values by your teammate. In general, interobserver reliability should exceed .70, with extensive training often increasing the r value to .80 or .90 levels.

ASSESSMENT QUESTIONS

Please address the following questions:

1. Which observation time yielded the higher r value?
2. Why are operational definitions important when seeking interobserver reliability?
3. Would a negative r value for interobserver reliability be likely? Why or why not?
4. Most likely, your interobserver reliability increased after you and your partner trained together. Why do think this increase occurred? If your interobserver reliability remained low, how might you increase it?

23

USING A CLASSIC MODEL OF STRESS TO TEACH SURVEY CONSTRUCTION AND ANALYSIS

Joseph A. Wister

The purpose of the activity presented in this chapter is to familiarize students with three important elements in survey development: (a) levels of measurement, (b) operational definitions, and (c) item design. In the activity, students develop their own survey using a classic organizational model of stress applied to the university setting.

CONCEPT

Surveys and questionnaires are used in a large number of psychology studies, so an understanding of survey construction is important for psychology students. Students learn better when they are engaged. Important factors in student engagement are the perceived relevance of the material to the students' lives and the degree of active and collaborative learning (Carini, Kuh, & Klein, 2006; Wister, 2002). The activity introduced in this chapter engages students in learning by having them develop their own survey using a classic organizational model of stress applied to the university setting. The students develop a survey, formulate hypotheses, collect data, and test their hypotheses while learning the nuances of survey construction.

MATERIALS NEEDED

Students should have access to Microsoft PowerPoint; a data analysis program (e.g., SPSS); and, if available, a survey distribution software package such as Qualtrics (https://www.qualtrics.com/) or SurveyMonkey (https://www.surveymonkey.com/).

INSTRUCTIONS

You should introduce students to the Institute of Social Research (ISR) model of stress (Caplan, Cobb, French, Harrison, & Pinneau, 1975; Hurrell & McLaney, 1988; Katz & Kahn, 1980), also known as the Michigan model. The ISR model was one of the first models of occupational stress, and it has served as the basis of much of the work in the area (see Mark & Smith, 2008). Strengths of the model in the context of this activity include its simplicity, generality, and heuristic value.

Introduction of the model typically takes one class period. As shown in Figure 23.1, the model begins with the *objective environment*, which is composed of variables that can be measured independently of the employee, such as number of hours of work per week. The *subjective environment* is the employee's appraisal of the objective environment. One way of viewing appraisal is the *job demands–resources model* (Bakker & Demerouti, 2007), which postulates that employees evaluate the demands placed on them within the context of resources available to them. If employees perceive that available resources are inadequate to meet demand, they experience threat, which is a precursor to strain.

---

http://dx.doi.org/10.1037/0000024-024

Activities for Teaching Statistics and Research Methods: A Guide for Psychology Instructors, J. R. Stowell and W. E. Addison (Editors)

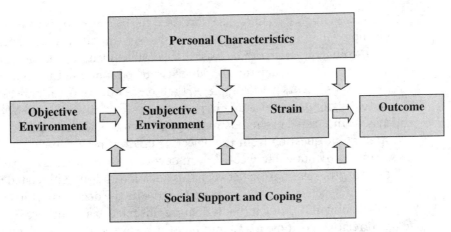

Figure 23.1. The Institute of Social Research model of stress.

Strain comprises the physiological, behavioral, and emotional responses to threat. Physiological responses include sympathetic nervous system arousal, such as an increase in blood pressure. Behavioral changes include a decrease in effort or concentration, and emotional changes include an increase in anxiety or a decrease in job satisfaction.

The last component of the model is *long-term outcome*. The focus of this component traditionally has been stress-related physical or mental health issues, such as cardiovascular disease or depression. For this activity, short-term behavioral outcomes such as decreased productivity can be used.

The model proposes that the four main components are moderated by two additional factors: (a) characteristics of the person, which include personality factors, skills, and abilities; and (b) social support, which includes both formal and informal support processes. I have added an additional moderating variable: coping mechanisms, which include both problem-focused and emotion-focused coping.

After introducing the students to the ISR model, divide them into groups, with each group assigned one of the components or moderating factors of the model. The students will have 1 week to investigate their topic and create PowerPoint presentations that summarize their research and describe how they might measure their component of the model within the college setting. For example, students assigned the topic of the objective environment may present research on attempts to measure outside stressors. The Daily Hassles Scale (Holm & Holroyd, 1992) and the Holmes–Rahe Life Stress Inventory (Holmes & Rahe, 1967) are examples of general surveys that measure the objective environment. The students should use these resources to generate a list of eight survey items specific to their college environment.

The student groups should present their research and survey questions to the class in 15-minute PowerPoint presentations. After the presentations are completed, instruct the students to take the surveys themselves, serving as a pilot group. The next class period is devoted to a full-class discussion of the concepts of the model and a revision of the survey items developed by each group. This activity is guided by the work of Rea and Parker (2014) and Willis and Lessler (1999) on survey construction (see a summary of questions in Appendix 23.1). At the end of this class there should be eight revised survey items for each component of the model.

The completed survey is submitted to the school's institutional review board and, on approval, is distributed to participants (e.g., members of the department's participant pool, volunteers from other classes). If possible, load the questionnaire on data collection software such as Qualtrics to reduce time and increase the accuracy of data entry.

After students have completed their data collection, they are assigned to new groups, and each group generates four hypotheses based on the ISR model. For example, one group may be interested in the relationship between outside activities and feelings of threat or strain. One question from the objective environment component of the survey may be "How many hours per week do you devote to extracurricular activities?" Students may examine the relationship between this question and an item related to the subjective environment, such as "On a 7-point scale, indicate the degree to which you feel overwhelmed by schoolwork." Students could analyze this relationship using a Pearson's r or Spearman's ρ correlation coefficient and interpret the results within the context of the model.

Each group should present their findings to the class in the form of a PowerPoint or poster presentation. The final stage of the project is a full-class reflection on and evaluation of the survey. This discussion can include the clarity of the questions, interpretation challenges, and suggestions for improvement. Often, the strengths and weaknesses of a survey become apparent only after the survey has been administered and the data interpreted.

ASSESSMENT

The mastery of the concepts presented in this activity can be evaluated by assignments during the course of the project. For example, students can be assessed on their ability to address the following items:

1. identify or create questions with different levels of measurement;
2. generate different operational definitions of a single concept (e.g., workload);
3. generate hypotheses that are clear, testable, and logical;
4. correctly implement the proper statistical analysis of their survey; and
5. present a clear and logical interpretation of their findings.

In addition, the quality of the items generated for the survey can be assessed by using the list of questions in Appendix 23.1 as a rubric.

DISCUSSION

When constructing the surveys, students can be directed to specific issues in methodology, such as incorporating items that make use of different levels of measurement—nominal, ordinal, interval, and ratio—as well as including Likert scales or other item types. The second element of survey construction that can be addressed is the distinction between variables that are directly observable, such as the number of credits a student is taking, and variables that have no direct sensory referent and can be defined only operationally, such as the degree to which a student feels overwhelmed by his or her courseload. Students can be asked to compare and contrast the same concept in the objective and subjective environments.

The design of survey items can also be addressed with this activity. Survey items should be clear, logical, and free from bias, and instructions should be easy to follow. After taking the surveys themselves and using the list of questions in Appendix 23.1, students can discuss the potential threats to these concepts and make appropriate changes to the survey.

The last part of the activity is to have students generate and test hypotheses using the ISR model and then present their results to the class. At the conclusion of the activity, students are asked to reflect on their experience. They are specifically asked what they would do differently, whether they would change any of the survey items, or add or eliminate items, and so son.

Although this activity is relatively elaborate and takes at least 3 to 4 weeks to complete, it involves students in the complete research enterprise, from theory to presentation of results. Students are exposed to the complexity of survey design and analysis in a manner that promotes critical thinking, fosters collaboration, and is relevant to their own experiences.

REFERENCES

Bakker, A. B., & Demerouti, E. (2007). The job demands–resources model: State of the art. *Journal of Managerial Psychology, 22,* 309–328. http://dx.doi.org/10.1108/02683940710733115

Caplan, R. D., Cobb, S., French, J. R., Harrison, R. D., & Pinneau, S. R. (1975). *Job demands and worker health: Main effects and occupational differences.* Washington, DC: U.S. Government Printing Office.

Carini, R. M., Kuh, G., & Klein, S. P. (2006). Student engagement and student learning: Testing the linkages research in higher education. *Research in Higher Education, 47,* 1–32. http://dx.doi.org/10.1007/s11162-005-8150-9

Holm, J. E., & Holroyd, K. A. (1992). The Daily Hassles Scale (Revised): Does it measure stress or symptoms? *Behavioral Assessment, 14,* 465–482.

Holmes, T. H., & Rahe, R. H. (1967). The Social Readjustment Rating Scale. *Journal of Psychosomatic Research, 11,* 213–218. http://dx.doi.org/10.1016/0022-3999(67)90010-4

Hurrell, J. J., Jr., & McLaney, M. A. (1988). Exposure to job stress—A new psychometric instrument. *Scandinavian Journal of Work, Environment & Health, 14*(1, Suppl. 1), 27–28.

Katz, D., & Kahn, R. L. (1980). *The social psychology of organizations* (2nd ed.). New York, NY: Wiley.

Mark, G. M., & Smith, A. P. (2008). Stress models: A review and suggested new direction. In J. Houdmont & S. Leka (Eds.), *Occupational health psychology: European perspectives on research, education, and practice* (Vol. 3, pp. 111–144). Nottingham, England: Nottingham University Press.

Rea, L. M., & Parker, R. A. (2014). *Designing and conducting survey research: A comprehensive guide* (4th ed.). San Francisco, CA: Jossey-Bass.

Willis, G. B., & Lessler, J. T. (1999). *Question Appraisal System QAS-99.* Rockville, MD: Research Triangle Institute. Retrieved from http://www.websm.org/uploadi/editor/1364216022Willis_Lessler_1999_QAS_99.pdf

Wister, J. A. (2002, June). *Active learning: The use of a student stress survey in a statistics class.* Poster presented at the 14th Annual Convention of the Association for Psychological Science, New Orleans, LA.

Appendix 23.1

Questions for Survey Evaluation

1. Are the instructions of the survey easy to follow?
2. Are the survey items easy to read?
3. Do any of the survey items lead the subject to a particular response?
4. Are any of the survey items confusing?
5. Are there questions that ask about two or more concepts but allow for only one answer (i.e., double-barreled questions)?
6. Are any of the survey items insensitive to issues of gender, race, ethnicity, sexual orientation, religion, or class?
7. Are any of the survey items too long?
8. Do the survey items contain jargon or highly technical wording that may be difficult to understand?
9. Is the order of the survey items logical?
10. Are there overlapping response options?
11. Is the survey too long?

24 USING CHILDHOOD MEMORIES TO DEMONSTRATE PRINCIPLES OF QUALITATIVE RESEARCH

Steven A. Meyers

The activity described in this chapter provides students with an engaging introduction to qualitative research by having them gather narrative descriptions of early childhood memories. Students practice coding these data by using established systems or by developing their own frameworks to detect themes and patterns. The activity's instructions allow for adjustments to its length and scope to reflect instructor preference or time constraints.

CONCEPT

Qualitative research methods significantly expand the analytic toolkit of psychology researchers, but this topic is not consistently included in undergraduate psychology research methods classes. Activities that allow students to use qualitative methods can be especially helpful because textbooks often lack detailed coverage, and this approach could otherwise be confusing to undergraduates because of its different assumptions and distinct strategies in comparison to quantitative research. A qualitative approach to research can be useful for psychology students to learn because it provides rich data, facilitates theory creation, structures the analysis of case studies, and even illustrates findings identified by quantitative studies in mixed-method designs.

Instructors can emphasize the following five overarching concepts when discussing qualitative research (cf. Flick, 2014): (a) Qualitative researchers often examine interview, narrative, or observational data rather than responses to questionnaires or experimental task performance; (b) their studies often involve smaller samples that are characterized by significant depth of analysis; (c) they develop categories to code data that capture important themes; and (d) they search for conceptual linkages and connections within the data after they are collected, instead of using a priori hypotheses and inferential statistics.

This activity illustrates these overarching concepts by having students collect qualitative data using detailed descriptions of early childhood memories (ECMs). Using methods instituted by Freud (1910/1990), ECM data are interesting, relevant, and easy to obtain. Students code these data using either an established system or grounded theory (Glaser & Strauss, 1967) to develop categories for detailed analysis.

MATERIALS NEEDED

The scope of this activity can be modified depending on instructor preference. In its limited implementation, students need access only to word processing software. For more thorough versions, required materials include digital voice recorders, shared access to computer files (i.e., cloud-based file storage), and qualitative analysis software.

http://dx.doi.org/10.1037/0000024-025
Activities for Teaching Statistics and Research Methods: A Guide for Psychology Instructors, J. R. Stowell and W. E. Addison (Editors)

Students gather qualitative data for this activity by either describing their own ECMs or by interviewing others. The following question elicits the initial description of an early childhood memory (Clark, 2002): "Think back to a long time ago when you were little, and try to recall one of your earliest memories, one of the first things that you can remember." The three follow-up questions are (a) "Is there anything else that you can recall in the memory?" (b) "What part do you remember most in the memory?" and (c) "What feelings do you remember having then?" To obtain additional data, each student must describe two more ECMs by answering the same questions. Students are likely to demonstrate repeating themes in a series of recollections, which mirrors the use of ECM tests when used in research and clinical settings (Mosak & Di Pietro, 2006).

There are two options for how the data can be recorded for this activity. The first option is that the ECM task can be presented as a written exercise to students, such that they write or type their answers to the prompt questions. The second option is for students to interview participants, record responses to the ECM questions, and then transcribe them. This approach is more labor intensive, but it is a closer approximation to how qualitative researchers obtain data because it allows for more detailed and natural responses.

In a brief implementation of this activity, each student serves as a sole participant, writes down his or her own ECMs, and subsequently analyzes these data. In a more extensive application, students obtain data from others (either inside or outside of the research methods class), pool ECM responses across multiple participants, and analyze data from this larger sample.

The most involved step of the activity is coding. This process allows students to organize, simplify, and capture important themes in the data. Students have the option of coding qualitative data using an established system or a manual. Although there are numerous scoring systems for ECMs, one that is well suited for this exercise is the Manaster–Perryman Manifest Content Early Recollections scoring manual (Manaster & Perryman, 1979) because it is intuitive and relatively easy to use. For each memory, students code the characters mentioned in the story (e.g., mother, father, siblings), the themes of the memory (e.g., birth of a sibling, death, illness or injury, punishment), the level of detail conveyed in the memory (e.g., attention to visual, auditory, and movement details), the setting of the story (e.g., school, hospital, at home, traveling), whether the person was active or passive in the memory, the level of control or responsibility that the respondent assumes for details in the story, and the predominant emotion or affect conveyed in the memory (e.g., positive, negative, or neutral tone).

Alternatively, you can design the activity so that students determine themes within the data by examining the responses, develop descriptive codes based on close reading, create a codebook, and then apply the codes to the ECM data to discern patterns of associations. This approach is consistent with grounded theory within qualitative research (Glaser & Strauss, 1967), which emphasizes that participants' experiences and perspectives should be given priority over preexisting theory or researchers' expectations.

A code sheet can facilitate analysis, especially if students are using an existing scoring system. Students can also code the responses with the assistance of word processing software. Character formatting (e.g., colors, bold, italics) or the Track Changes function in Word can indicate coding or comments. La Pelle (2004) described how to use tables so that one column contains the source data and adjacent columns can be used for the codes that subsequently can be sorted for retrieval and analysis (see Table 24.1 for a

Table 24.1 *Data Table Excerpt*

Part No.	Male (M) or female (F)	Age in story	Segment No.	Response	Codes
1	M	3	1	It was my third birthday party. I remember being excited and I was looking forward to eating my birthday cake.[1] Several of my friends were there who lived in my neighborhood. My mom, dad, and sister were there as well. As they started to sing "Happy Birthday," a clown came in and started to sing along. I guess he was there to entertain all of the kids later in the party. I didn't expect the clown to be there and it scared me.[2] I remember that I started to cry because I was so freaked out. I was upset for the rest of the party.[3] *"Is there anything else that you can recall in the memory?"*	[1]Positive emotion/excited/event [2]Negative emotion/fear/stranger [3]Negative emotion/sad/event
1	M	3	2	I remember that my mom tried really hard to comfort me.[4] She assured me that the clown was friendly, and that he was going to do cool tricks with balloons.[4] I ran off to my room and stayed there until the clown left the party.	[4]Positive emotion/comfort/parent

simplified, illustrative analysis). Students ultimately are asked to find text illustrations of the codes, discern patterns among the codes, and determine how themes vary with other constructs of interest (e.g., participant gender, age at the time of the memory). You can also use dedicated qualitative data analysis software to assist with more complex analyses (see the Additional Resources section at the end of this chapter).

ASSESSMENT

You can use simple assessment strategies, such as measures of student satisfaction with the activity or a self-assessment of its effectiveness. Another option is to include administering an objective test of knowledge about qualitative research before and after the activity has been completed. Additional assessment opportunities include evaluating students' learning products from this activity (e.g., coding manual, summary of the results and analyses, poster presentation) and having students design a novel qualitative study so that the transfer of training can be assessed.

DISCUSSION

Students generally find this activity enjoyable because the content is personalized and connected to their experiences. The ECM data are easily understandable and allow students to gain an awareness about themselves and their peers. Nevertheless, there can be potential barriers. For example, some students can have difficulty recalling specific ECMs. This could reflect ambivalence about the task, or it may reflect actual memory limitations. Although in clinical and research settings participants are asked to recall an incident in which they were younger than age 9 years, the activity will still be valuable if they recall a more recent event.

In addition, some students likely experienced highly distressing events or trauma during their childhoods. Because of this, you can opt to modify the instructions to include a disclaimer about potential distress, or you can state that students can selectively choose or edit any memory that they share. This concern, as well as the previous one, can be mitigated by having students analyze data from peers rather than requiring all students to generate or examine their own data. Although this alteration may reduce the self-awareness generated by the activity, exchanging or pooling data will likely reduce any self-serving biases as well.

Finally, this activity allows instructors to use ECMs as a means to connect with related topics within the psychology curriculum. These topics include long-term memory limitations and childhood amnesia, stages of child development that parallel the time frame of the ECMs, and how psychology theorists and researchers have viewed childhood experiences as determinants of personality and/or psychopathology.

ADDITIONAL RESOURCES

Instructors can share examples of published qualitative research from peer-reviewed journals with students. Exemplary studies can be found in issues of *Qualitative Psychology*, *Qualitative Research in Psychology*, and *Feminism & Psychology*.

Instructors who want to introduce students to sophisticated software for qualitative data management as part of this activity have several choices. Some are open source or have free trial versions. Recommended options include ATLAS.ti (http://atlasti. com/qualitative-data-analysis-software/); NVivo (http://www.qsrinternational.com/ products_nvivo.aspx); QDA Miner (http://provalisresearch.com/products/qualitative-data-analysis-software/); and Weft QDA (http://www.pressure.to/qda/).

REFERENCES

Clark, A. J. (2002). *Early recollections: Theory and practice in counseling and psychotherapy*. New York, NY: Brunner-Routledge.

Flick, U. (2014). *An introduction to qualitative research* (5th ed.). Thousand Oaks, CA: Sage.

Freud, S. (1990). *Leonardo da Vinci and a memory of his childhood*. New York, NY: Norton. (Original work published 1910)

Glaser, B. G., & Strauss, A. L. (1967). *The discovery of grounded theory: Strategies for qualitative research*. New York, NY: Aldine.

La Pelle, N. (2004). Simplifying qualitative data analysis using general purpose software tools. *Field Methods, 16*, 85–108.

Manaster, G. J., & Perryman, T. B. (1979). Manaster–Perryman Manifest Content Early Recollections scoring manual. In H. A. Olson (Ed.), *Early recollections: Their use in diagnosis and psychotherapy* (pp. 347–353). Springfield, IL: Charles C Thomas.

Mosak, H. H., & Di Pietro, R. (2006). *Early recollections: Interpretive method and application*. New York, NY: Routledge.

25 USING A PEER-WRITING WORKSHOP TO HELP STUDENTS LEARN AMERICAN PSYCHOLOGICAL ASSOCIATION STYLE

Dana S. Dunn

Learning to write American Psychological Association (APA) Style papers in research methods classes can be challenging for many students, just as teaching APA Style sometimes can be a struggle for instructors. Instituting a writing workshop for peer partners to share draft sections of APA Style papers ensures regular, incremental progress on papers and continual feedback that improves the quality of students' papers.

CONCEPT

Writing workshops for undergraduates are predicated on the idea that sharing writing and constructively responding to the writing of others improves students' own writing skills (Elbow & Belanoff, 1999; Hafer, 2014). In addition, writing practice is linked to improved writing quality (Arum & Roksa, 2011). Research methods courses in psychology usually have some empirical component, which requires students to collect and analyze data and then write up their findings in the form of an APA Style paper.

I frequently teach a two-semester sequenced methods and statistics course in which students design and execute their own experiments. To help students learn APA Style, I devote several class sessions during each semester to a writing workshop wherein peer partners read and comment on multiple drafts of a research proposal (first semester) and drafts of a final research paper (second semester). The workshop approach is flexible because it can also be used in one-semester methods courses or in topical seminars that require APA literature review–style papers.

MATERIALS NEEDED

Students must have access to a physical or digital copy of the *Publication Manual of the American Psychological Association* (APA, 2010) or another text that provides detailed instructions and examples of APA Style writing and citation referencing (e.g., Beins, 2012; Dunn, 2011). Providing students with a good model of a complete APA Style manuscript is helpful (e.g., a student paper from an earlier iteration of your course or a mock-up you created), as is a grading rubric. Alternatively, you might have the class read and discuss a simple, well-written APA Style paper you select from the published literature to serve as a model. Students should also have a copy of the Writing Workshop Guidelines handout (see Appendix 25.1) that you discuss with them in detail. These guidelines help foster an open climate for commentary and criticism

http://dx.doi.org/10.1037/0000024-026

Activities for Teaching Statistics and Research Methods: A Guide for Psychology Instructors, J. R. Stowell and W. E. Addison (Editors)

among peers, one that unfolds gradually as they become accustomed to giving and receiving feedback. Beyond these materials, they will need an ample supply of paper for printing draft sections of their papers that they will swap with their peers during the workshop. Bringing drafts to class on laptops or tablets is a possibility as long as editing software is also available on the device (e.g., Track Changes in Microsoft Word).

You should assign students to writing pairs for the semester (if you have an odd number of students, a trio is fine). Some instructors like to pair students who have stronger writing skills with those who are developing writers. Explain that the class will have a writing workshop component wherein once or twice each week students will be expected to bring a draft of one section of their developing APA Style research paper (e.g., Abstract, Introduction, Method, Results, Discussion). Peer pairs will swap section drafts, then read, evaluate, and provide written feedback, followed by verbal exchange of feedback. Writing workshops usually last about 35 minutes during the early part of a semester and then can take 70 minutes toward the end of the semester. The workshops are usually held for about 6 weeks each semester. (Some instructors may have less time available, which means the workshop can occur late in the semester, a week or two before the final paper is due—any peer feedback is apt to be more helpful than no feedback at all.)

Although one goal of the workshop process is to identify weaknesses in writing or deviations from APA Style, students should also be encouraged to identify the positive strengths they see in the work of their peers. Thus, the Writing Workshop Guidelines should be kept close at hand and referred to often until students have become comfortable and familiar with the peer review process. The main goal is to establish a sense of trust in the classroom so that everyone, including the instructor, can learn to improve their writing on the basis of feedback from peer readers. As the instructor, you can model how to be a good peer reviewer by reading and commenting on one or two student papers during each workshop session or filling in as a peer when a partner is absent. You can also take a moment or two toward the end of a workshop session to share the APA Style strengths and weaknesses you noted in the papers you reviewed that day.

Toward the end of each workshop session, I often ask the class to share some of the positive qualities they see in the developing papers. Some students will share a paragraph they believe is well written, whereas others will highlight a clear thesis statement or a good description of a piece of published research. Because of the trust established in the class, students are also encouraged to diplomatically describe writing challenges they see in their peers' papers. For example, some peers will identify the need to create clearer transitions between paragraphs or to provide concrete, behaviorally based examples that illustrate a theory, research finding, or point being made.

If your institution has a writing center, you may consider requiring your students to make an appointment and to take a draft of their paper there for comments from an additional reviewer. The class period before the final paper is due should be set aside for peers to proofread each other's penultimate version with special emphasis on APA Style issues (e.g., placement of the running head, accuracy of citations in the References).

ASSESSMENT   There are a variety of ways to assess how well students are learning APA Style as a result of the workshop experience. First, students can be asked to maintain a portfolio containing all the drafts of their paper. Toward the end of the semester, they can review the contents of the portfolio and write a self-reflection on how their prowess at writing an APA Style paper has changed. You can then review the portfolios as well as students' self-reflections on their writing. Second, as previously noted, you can provide students with a grading rubric that you intend to use when evaluating their final papers. The rubric can help students craft quality APA Style papers while affording you a more objective way to grade their final work. Third, peer partners can be asked to write brief reflections about the quality of each other's APA Style writing on the basis of the workshop experience. Fourth, after receiving their graded final papers, students can be asked to review them in light of your comments and to write a reflection on how they can improve the writing of their next APA Style paper.

DISCUSSION   Research methods instructors often complain about the quality of the APA Style papers they receive. Coupled with this complaint is the realization that feedback late in the paper process (i.e., after it is graded) does little to help students learn to write using APA Style or to produce clear prose. The ongoing, incremental nature of peer feedback encourages students to work consistently to produce drafts and to revise what they have written while also attending to APA Style considerations. In theory, the workshop process imparts good writing habits more generally, including drafting sections one at a time, seeking constructive comments on rough drafts, and rewriting.

Naturally, both my students and I do encounter some problems with the Writing Workshop. For their part, students need to be reminded that revising a draft is not simply making a few changes based on peer comments; instead, they must learn to revise the entire draft, not just a few sentences or some punctuation errors. They must also learn to balance writing and revising drafts with the other requirements of the research methods and statistics course, such as keeping up on reading, doing problem sets as homework, and studying for examinations. For my part, I need to remember that writing—good writing—takes time and that my students have at least three other courses to juggle along with my rather intensive course. To that end, I regularly offer praise and encouragement during the workshop sessions and emphasize how important writing is (or should be) to their undergraduate educational experience.

I also recognize that instructors who have classes with large enrollments may need to modify the workshop approach. Some may want to simply set aside one or two class meetings each semester in which students are encouraged to bring in a rough draft to share with whomever is sitting next to them. Alternatively, a voluntary, 1-hour, once-a-semester, outside-of-class meeting can be made available to students who want to share their working papers with peers and their professor.

Learning to communicate effectively and efficiently through written prose arguably is one of the most important learning outcomes of an undergraduate education, whether in psychology or another discipline. Requiring students to think and to write using APA Style provides educational benefits, including imparting the "fundamental attitudes and values of psychologists" (Madigan, Johnson, & Linton, 1995, p. 428). By learning psychology's language conventions, students, in turn, implicitly come to endorse the discipline's empirical perspective.

REFERENCES American Psychological Association. (2010). *Publication manual of the American Psychological Association* (6th ed.). Washington, DC: Author.

Arum, R., & Roksa, J. (2011). *Academically adrift: Limited learning on college campuses.* Chicago, IL: University of Chicago Press.

Beins, B. C. (2012). *APA Style simplified: Writing in psychology, education, nursing, and sociology.* Malden, MA: Wiley-Blackwell.

Dunn, D. S. (2011). *A short guide to writing about psychology* (3rd ed.). New York, NY: Longman.

Elbow, P., & Belanoff, P. (1999). *A community of writers: A workshop course in writing* (3rd ed.). New York, NY: McGraw-Hill.

Hafer, G. R. (2014). *Embracing writing: Ways to teach reluctant writers in any college course.* San Francisco, CA: Jossey-Bass.

Madigan, R., Johnson, S., & Linton, P. (1995). The language of psychology: APA Style as epistemology. *American Psychologist, 50,* 428–436. http://dx.doi.org/10.1037/0003-066X.50.6.428

Appendix 25.1

Writing Workshop Guidelines

1. Write constructive and helpful comments on your peer's draft or paper. Be specific. You can be critical, but your goal is not to upset the author. You want to help the author improve the paper.
2. Focus first on the content of the paper. Grammar, punctuation, and American Psychological Association Style issues are best addressed after you have provided comments on the content.
3. Look for the main point of the writing. If you are unsure what it is, ask the author about it after you finish reading and commenting on the draft.
4. What do you want or need to know that is not currently discussed in the draft? How can you help the author add it?
5. How can the draft be improved? Are there parts of the draft that need clarification or illustrative examples?
6. What part of the paper made the most sense to you? The least?
7. Identify the clearest sentence in the draft. Then locate the sentence that is hardest to understand.
8. Should any parts of the draft be edited out because they do not support the thesis?
9. Does the draft end well?
10. Tell the author what you liked best about the paper.

1. Readers are always right. As an author, try not to be defensive about the observations or suggestions your peer reader makes; instead, accept any comments or suggestions graciously. Reflect on how some or all of them might improve your work for readers.
2. In the end, this is still your paper. You do not have to make all the changes a reader recommends, but you should still consider them as you write the next draft.
3. Ask your peer reader what section of the draft or paper was clear, as well as what parts were more difficult to understand.
4. Ask whether the transitions between paragraphs make sense or if they are abrupt.
5. Ask your peer reader whether there are any specific changes that could improve the draft.
6. Thank your peer reader for providing comments.

INDEX

ABOUT THE EDITORS

Jeffrey R. Stowell, PhD, earned his doctoral degree in psychobiology from The Ohio State University. He is a professor and the assistant chair of the psychology department at Eastern Illinois University (EIU), where he teaches courses in biological psychology, sensation and perception, learning, and introductory psychology. He has published articles in *Teaching of Psychology*, *Scholarship of Teaching and Learning in Psychology*, and other teaching-related journals on the use of technology in teaching. He presents regularly at regional psychology conferences and mentors undergraduate and graduate student research. He participated in the 2008 National Conference on Undergraduate Education in Psychology: A Blueprint for the Future of the Discipline. He received the Society for Teaching of Psychology's Early Career Teaching Award and served as the society's Internet editor for 8 years. At EIU, Dr. Stowell has earned the honors of Professor Laureate and Distinguished Honors Faculty Award; he is a three-time winner of the Psi Chi Chapter Faculty of the Year Award and has received the College of Sciences' highest awards in three different areas (teaching, research, and service).

William E. Addison, PhD, is a professor in the Psychology Department at Eastern Illinois University (EIU), where he has regularly taught courses in statistics and research methods. He is a Fellow and former president of the Society for the Teaching of Psychology, Division 2 of the American Psychological Association, and he is a charter Fellow of the Midwestern Psychological Association. He has served as a consulting editor and reviewer for the journal *Teaching of Psychology*, as a member of the GRE Psychology Test Development Committee, and as a faculty consultant for the annual Advanced Placement Exam in Psychology. He participated in the 1999 National Forum on Psychology Partnerships and the 2008 National Conference on Undergraduate Education in Psychology: A Blueprint for the Future of the Discipline. Dr. Addison presents regularly at annual meetings of the American Psychological Association and the Midwestern Psychological Association and at the Midwest Institute for Students and Teachers of Psychology. His publications include teaching-oriented articles in *Teaching of Psychology* and the *College Student Journal*. He has received a number of awards for his teaching, including the EIU Distinguished Faculty Award and the EIU Distinguished Honors Faculty Award.